T0222765

Springer Undergraduate Mathematics Series

Springer

London
Berlin
Heidelberg
New York
Barcelona
Budapest
Hong Kong
Milan
Paris
Santa Clara
Singapore
Tokyo

Other books in this series

D.L. Johnson

Elements of Logic via Numbers and Sets

 Springer

D.L. Johnson, BSc, MSc, PhD
Department of Mathematics, University of Nottingham,
University Park, Nottingham NG7 2RD, UK

Cover illustration elements reproduced by kind permission of:

Aptech Systems, Inc., Publishers of the GAUSS Mathematical and Statistical System, 23804 S.E. Kent-Kangley Road, Maple Valley, WA 98038, USA. Tel: (206) 432 - 7855 Fax (206) 432 - 7832 email: info@aptech.com URL: www.aptech.com

American Statistical Association: Chance Vol 8 No 1, 1995 article by KS and KW Heiner 'Tree Rings of the Northern Shawangunks' page 32 fig 2

Springer-Verlag: Mathematica in Education and Research Vol 4 Issue 3 1995 article by Roman E Maeder, Beatrice Amrhein and Oliver Gloor 'Illustrated Mathematics: Visualization of Mathematical Objects' page 9 fig 11, originally published as a CD ROM 'Illustrated Mathematics' by TELOS: ISBN 0-387-14222-3, german edition by Birkhauser: ISBN 3-7643-5100-4.

Mathematica in Education and Research Vol 4 Issue 3 1995 article by Richard J Gaylord and Kazume Nishidate 'Traffic Engineering with Cellular Automata' page 35 fig 2. Mathematica in Education and Research Vol 5 Issue 2 1996 article by Michael Trott 'The Implicitization of a Trefoil Knot' page 14.

Mathematica in Education and Research Vol 5 Issue 2 1996 article by Lee de Cola 'Coins, Trees, Bars and Bells: Simulation of the Binomial Process page 19 fig 3. Mathematica in Education and Research Vol 5 Issue 2 1996 article by Richard Gaylord and Kazume Nishidate 'Contagious Spreading' page 33 fig 1. Mathematica in Education and Research Vol 5 Issue 2 1996 article by Joe Buhler and Stan Wagon 'Secrets of the Madelung Constant' page 50 fig 1.

ISBN 3-540-76123-3 Springer-Verlag Berlin Heidelberg New York

British Library Cataloguing in Publication Data
Johnson, David Lawrence
Elements of logic via numbers and sets. - (Springer undergraduate mathematics series)
    1.Number theory 2.Logic
    I.Title
    512.7
ISBN 3540761233

Library of Congress Cataloging-in-Publication Data
Johnson, D.L.
    Elements of logic via numbers and sets / D.L. Johnson
      p. cm. - (Springer undergraduate mathematics series)
    Includes bibliographical references (p. 165) and index.
    ISBN 3-540-76123-3 (Berlin: pbk.: acid-free paper)
    1. Logic, Symbolic and mathematical. I. Title. II. Series.
QA9.J63 1998   97-28662
511.3-dc21   CIP

Typeset by Focal Image, London
Printed and bound at the Athenæum Press Ltd., Gateshead, Tyne & Wear
12/3830-5432 Printed on acid-free paper SPIN 10831990

For Grace

Contents

Introduction ix

1. **Numbers** 1
   1.1 Arithmetic Progressions 1
   1.2 Proof by Contradiction 5
   1.3 Proof by Contraposition 8
   1.4 Proof by Induction 10
   1.5 Inductive Definition 19
   1.6 The Well-ordering Principle 27

2. **Logic** 35
   2.1 Propositions 35
   2.2 Truth Tables 39
   2.3 Syllogisms 43
   2.4 Quantifiers 48

3. **Sets** 53
   3.1 Introduction 54
   3.2 Operations 58
   3.3 Laws 62
   3.4 The Power Set 65

4. **Relations** 71
   4.1 Equivalence Relations 72
   4.2 Congruences 75
   4.3 Number Systems 79
   4.4 Orderings 85

5. Maps 89
  5.1 Terminology and Notation 89
  5.2 Examples... 94
  5.3 Injections, Surjections and Bijections...................... 99
  5.4 Peano's Axioms 105

6. Cardinal Numbers 113
  6.1 Cardinal Arithmetic 114
  6.2 The Cantor–Schroeder–Bernstein theorem 118
  6.3 Countable Sets 121
  6.4 Uncountable Sets 126

Solutions to Exercises 131

Guide to the Literature....................................... 163

Bibliography... 165

Dramatis Personae 167

Index 171

Introduction

This book is based on a module given to first-year undergraduates at the University of Nottingham with the aim of bridging the gap between school and university in mathematics. This is not so much a gap in the substance or material content of the subject but more a change in attitude and approach. In pure mathematics for example, rather than memorize a formula and be able to apply it, we want to understand that formula and be able to prove it. Physical intuition does not constitute a proof and neither does accumulated statistical evidence; what is required is a formal logical process. Since logic can sometimes appear rather a dry subject, we take as an underlying theme the concept of "number", which not only provides a rich source of illustrations but also helps to lay the foundations for many areas of more advanced mathematical study.

We begin in Chapter 1 with a survey of useful facts about numbers that are more or less familiar, such as the binomial theorem and Euclid's algorithm, respectively, giving formal proofs of different kinds, notably proof by induction. It will become increasingly evident that to gain a better understanding of what happens in the course of a proof we need a systematic language or framework within which to develop the ideas involved. Fortunately, one exists, and forms the subject of the next chapter.

The first half of Chapter 2 looks at propositional calculus, which treats propositions (statements that are either true of false) and the relations between them (such as implication) as mathematical objects. Ways of combining these and of operating on them, familiar from our knowledge of language, are put into formal shape governed by precise rules. The second half of Chapter 2 studies syllogisms, which are logical arguments, or processes of deduction, of the simplest kind. They thus form an ideal model or pattern for all forms of logical inference, and also serve to introduce the deceptively simple but crucially important idea of quantification.

The core of the book is Chapter 3, where the development of set theory closely parallels that of logic in the previous chapter. It also lays the foundations for important results in the later chapters and introduces the terminology and notation that comprise the language of modern mathematics.

Two important types of relation are studied in Chapter 4: the notion of ordering forms a familiar and fundamental element of structure in most number systems, and the seemingly artificial concept of equivalence relation. The latter finds application in many areas of advanced mathematics, especially in the construction of number systems and in abstract algebra, where the key result Theorem 4.1 plays a decisive role.

The elementary concept of function is generalized to the more abstract idea of map in Chapter 5, where many more or less familiar examples are given. A study of the basic properties of maps leads, among other things, to our second key result of ubiquitous application, Theorem 5.6.

The revolutionary ideas of Cantor described in Chapter 6 bring us very nearly into the present century. In addition to proving our third key result, Theorem 6.6, we make good the claim that the concept of set is more fundamental than that of number, and the axiomatic development of number systems reaches its crowning glory in the construction of the real numbers. The diligent reader will be rewarded with at least a glimpse of the meaning of the infinite.

It is a pleasure to acknowledge my gratitude to Springer-Verlag, and especially to Susan Hezlet, for their courteous and efficient handling of all matters connected with the production of the book.

D.L.J.

1
Numbers

Historically, mathematics came into being to serve two purposes, counting and measuring. Both of these required the use of *numbers*, the positive integers $\mathbb{N}$ and the real numbers $\mathbb{R}$, respectively. The need to solve equations such as

$$2x = 3, \qquad x + 4 = 0, \qquad x^2 + 1 = 0$$

subsequently led to the appearance of more sophisticated number systems like the rational numbers $\mathbb{Q}$, the integers $\mathbb{Z}$, and the complex numbers $\mathbb{C}$. We are chiefly concerned here with properties of the positive integers and, at the same time, the means by which such properties are established. This revolves around the concept of a mathematical *proof*, of which we give examples of four kinds, finishing up with the most important for $\mathbb{N}$, proof by induction.

1.1 Arithmetic Progressions

Definition 1.1

An **arithmetic progression** is a sequence of numbers

$$a_0, a_1, a_2, \ldots, a_k, \ldots$$

whose consecutive terms differ by a constant called the **common difference**

$$d = a_1 - a_0 = a_2 - a_1 = \cdots = a_k - a_{k-1} = \cdots. \tag{1.1}$$

Thus, if the **first term** $a_0 = a$ say, then

$$a_1 = a + d, \qquad a_2 = a + 2d; \qquad a_k = a + kd \qquad (1.2)$$

in general. We seek a formula for the sum of the first n terms of this sequence,

$$s_n = a_0 + a_1 + \cdots + a_{n-1}, \qquad (1.3)$$

which is usually arrived at in the following way. Compare (1.3) with the same sum "written backwards",

$$s_n = a_{n-1} + a_{n-2} + \cdots + a_0 \qquad (1.4)$$

and observe that the n pairs of terms aligned vertically all have the same sum,

$$a_{k-1} + a_{n-k} = a + (k-1)d + a + (n-k)d = 2a + (n-1)d, \qquad (1.5)$$

$k = 1, \ldots, n$, so that n times this number is equal to twice the desired sum, that is,

$$2s_n = n(2a + (n-1)d). \qquad (1.6)$$

Theorem 1.1

The sum s_n of the first n terms of the arithmetic progression with first term a and common difference d is given by the formula

$$s_n = na + \tfrac{1}{2}n(n-1)d. \qquad (1.7)$$

□

While this process of derivation has a number of points in its favour, namely, it

(a) contains the key idea—reversing the sum,

(b) is fairly convincing—you believe the result,

(c) leads to the right answer—formula (1.7) is correct,

there are nevertheless some shortcomings, such as:

(i) the dots $\cdots$ in (1.1), (1.3), (1.4) are not defined precisely,

(ii) the formula $a_k = a + kd$ in (1.2) has not been proved, although it is almost "obvious",

(iii) tacit assumptions are made in the manipulations leading successively to (1.5), (1.6), (1.7).

We attempt to remedy each of these failings in turn.

First of all, the dots $\cdots$ in (1.1) will be eliminated if we replace all these equations by the single assertion

for every positive integer $k, a_k - a_{k-1} = d$.

Next, the dots $\cdots$ in (1.3) and (1.4) can be eliminated by passing to so-called Σ-notation, pronounced "sigma-notation", when they become

$$s_n = \sum_{k=0}^{n-1} a_k, \qquad s_n = \sum_{k=1}^{n} a_{n-k}. \tag{1.8}$$

In each of these equations, the right-hand side is evaluated by substituting the indicated values of k in the given expression and adding the resulting numbers. The fact that these two apparently different expressions for s_n are formally equal is a consequence of various rules of manipulation in Σ-notation like those in the exercises below. These rules are in turn the consequences of fundamental laws of arithmetic referred to after the next paragraph.

Turning to (ii), the general formula in (1.2) can be proved either by induction or by using the Σ-rules just mentioned. Since the latter are also proved by induction, the remedy for this failing must be deferred until Section 1.4.

Finally, the tacit assumptions referred to in (iii) are all of the type introduced as axioms (see below) in elementary algebra and more properly called **fundamental laws of arithmetic**. Perhaps the most prominent of these are the **commutative laws**:

$$a + b = b + a \quad \text{and} \quad ab = ba, \tag{1.9}$$

of addition and multiplication, respectively. The term "law" means that these equations hold *for all* a, b, c in $\mathbb{N}$. They generalize to sums and products of any number n of terms, and thus prove that the right-hand sides of (1.3) and (1.4) are equal.

Next, rather more subtly, notice that the right-hand side of (1.3) contains not only n terms but also $n - 1$ *operations* (of addition), and, just as the former can be written in any order, the latter can be performed in any order. This is a consequence of the **associative laws**:

$$(a + b) + c = a + (b + c) \quad \text{and} \quad (ab)c = a(bc), \tag{1.10}$$

for all a, b, c in $\mathbb{N}$, of addition and multiplication, respectively.

Finally, in (1.5) and (1.7), we have made use of the **distributive laws**:

$$a(b + c) = ab + ac, \qquad (a + b)c = ac + bc, \tag{1.11}$$

valid for all a, b, c in $\mathbb{N}$. These two laws relate addition to multiplication and, like the commutative and associative laws, are commonly assumed and used without explicit reference. Note that when multiplication is commutative, each of the two distributive laws is a consequence of the other.

We have now assembled most of the axioms for a particularly important type of number system. To get the others, observe that the integers $\mathbb{Z}$ have an **additive identity** 0 and a **multiplicative identity** 1:

$$a + 0 = a, \qquad a1 = a, \tag{1.12}$$

for all integers a, and that every integer a has an **additive inverse** $-a$:

$$a + (-a) = 0. \tag{1.13}$$

A system closed under addition and multiplication for which (1.9)–(1.13) hold is called a **commutative ring-with-identity**. Removing the second equations in (1.9) and (1.12), we simply get a **ring**. Adjoining to (1.9)–(1.13) the extra axiom that there are no **divisors of zero**:

$$\text{if} \quad a \neq 0 \quad \text{and} \quad b \neq 0, \quad \text{then} \quad ab \neq 0, \tag{1.14}$$

gives an **integral domain**. The existence of **multiplicative inverses**:

$$\text{if} \quad a \neq 0, \quad \text{there is an } a^{-1} \text{ with } aa^{-1} = 1, \tag{1.15}$$

gives a **field**.

Thus, the integers $\mathbb{Z}$ form an integral domain but not a field. The rational numbers $\mathbb{Q}$, the real numbers $\mathbb{R}$, and the complex numbers $\mathbb{C}$ are all fields.

EXERCISES

Convince yourself that the following ten Σ-rules are all valid.

1.1 $\sum_{k=1}^{n} a = na$.

1.2 $a \sum_{k=1}^{n} a_k = \sum_{k=1}^{n} a a_k$.

1.3 $\sum_{k=1}^{n} a_k + \sum_{k=1}^{n} b_k = \sum_{k=1}^{n} (a_k + b_k)$.

1.4 $\sum_{k=1}^{n} a_k = \sum_{i=1}^{n} a_i$.

1.5 $\sum_{k=1}^{n} a_k = \sum_{k=0}^{n-1} a_{k+1}$.

1.6 $\sum_{k=0}^{n-1} a_k = \sum_{k=1}^{n} a_{n-k}$.

1.7 $\sum_{k=1}^{n} a_k = \sum_{k=1}^{n-1} a_k + a_n$.

1.8 $\left(\sum_{i=1}^{m} a_i \right) \left(\sum_{j=1}^{n} b_j \right) = \sum_{i=1}^{m} \left(\sum_{j=1}^{n} a_i b_j \right)$.

1.9 $\sum_{i=0}^{n-1}\left(\sum_{j=i+1}^{n} a_i b_j\right) = \sum_{j=1}^{n}\left(\sum_{i=0}^{j-1} a_i b_j\right)$.

1.10 $\left(\sum_{k=0}^{n} a_k\right)^2 = \sum_{k=0}^{n} a_k^2 + 2\sum_{i=0}^{n-1}\left(\sum_{j=i+1}^{n} a_i a_j\right)$.

1.11 Use these results to put together a more formal proof of formula (1.7).

1.12 The number of ways of ordering the n terms in the sum

$$s_n = \sum_{k=1}^{n} a_k$$

is, almost by definition, equal to $n!$, pronounced "n factorial". On the other hand, curiously, the number c_n of ways of ordering the $n-1$ operations in s_n, that is, the number of different bracketings, is not equal to $(n-1)!$ for $n \geq 4$. Calculate the value of c_n for some small values of n, and see what you can deduce about these numbers (called the **Catalan numbers**).

1.13 Prove that every field is an integral domain.

1.14 Prove that every finite integral domain is a field.

1.2 Proof by Contradiction

Suppose you wish to prove that a certain (mathematical) statement P is true. One way of doing this is to prove that P is not false, and this may be done as follows. Let $\sim P$ stand for the statement "P is false"; $\sim P$ is called the **negation** of P and pronounced "not P". Suppose you can deduce from $\sim P$ that a certain statement Q is both true and false; such a situation flies in the face of reason and a fundamental law of logic, and is called a **contradiction**. You can then conclude that your assumption $\sim P$ is wrong, that is, $\sim P$ is false, which is the same as saying that P is true. This method of proof is called "reductio an absurdum" in Euclidean geometry. It is more commonly known as **proof by contradiction**, and we shall give a more formal justification for its validity in Chapter 2. The example of this type of proof given below concerns numbers and requires a familiar definition.

Definition 1.2

A real number r is called **rational** if some non-zero integral multiple of r is an integer: $br = a$ for some integers a and b with $b \neq 0$, and then we write

$r = a/b$, pronounced "a over b". The expression a/b is called a **fraction** with **numerator** a and **denominator** b.

Examples of fractions are

$$1/2, \qquad 0/5, \qquad -2/3, \qquad 6/1, \qquad 3/6, \qquad 2/-3. \qquad (1.16)$$

These fractions all represent rational numbers, and the main reason why the treatment of fractions in elementary school is so difficult is that this representation is *not unique*, that is, different fractions can represent the same rational number. This is inherent in the definition, for the equation $br = a$ can be multiplied by any non-zero integer m to give the equivalent equation $mbr = ma$, so that the fractions ma/mb all represent the same rational number r. The way round this is to choose the "simplest" one, namely the one where

the denominator is the *smallest* positive integer multiple

of r to be an integer. $\qquad (1.17)$

In this case a and b can have no common factor (other than ± 1) and a/b is said to be **in lowest terms**.

The Ancient Greeks believed at first that all real numbers are rational. Imagine their horror when, as a consequence of Pythagoras' celebrated theorem, an irrational number appeared on the scene: the hypotenuse h of a right triangle with shorter sides both of length 1 satisfies the equation $h^2 = 2$; write $h = \sqrt{2}$, pronounced "root-two".

Theorem 1.2 (Euclid)

The number $h = \sqrt{2}$ is irrational.

Proof

Assume (for a contradiction) that the statement of Theorem 1.2 is false, that is, that h is rational. Then we can write $h = a/b$ in lowest terms, where a and b are integers. Then

$$h^2 = a^2/b^2 = 2,$$

whence $a^2 = 2b^2$. So a^2 is even and thus so is $a : a = 2c$, say. But then

$$4c^2 = a^2 = 2b^2,$$

whence $b^2 = 2c^2$. So b^2 is even and thus so is $b : b = 2d$, say. But then

$$h = a/b = 2c/2d = c/d,$$

which contradicts the fact that a/b is in lowest terms. We conclude that our original assumption, that h is rational, is false. Therefore, $h = \sqrt{2}$ is irrational, as required. $\square$

In this proof, we used the statements

$$
\begin{aligned}
P &: \quad \sqrt{2} \text{ is irrational,} \\
\sim P &: \quad \sqrt{2} \text{ is rational,} \\
Q &: \quad a/b \text{ is in lowest terms,} \\
\sim Q &: \quad a/b \text{ is not in lowest terms.}
\end{aligned}
$$

We already know that Q is true, by (1.17). We then deduced $\sim Q$ from $\sim P$. The contradiction Q and $\sim Q$ enables us to conclude P.

EXERCISES

1.15 Write the fractions given in (1.16) in lowest terms.

1.16 Convince yourself that the representation of a rational number satisfying (1.17) is unique.

1.17 Let n be an integer. Prove that the remainder on division of n^2 by 4 is 0 or 1 according as n is even or odd. What remainders are possible when n^2 is divided by 8?

1.18 Let c be a real number satisfying $c^3 = 5$; write $c = \sqrt[3]{5}$, the "cube-root of five". Prove that c is irrational.

1.19 Convince yourself that statements P and $\sim \sim P$ are the same. Suppose that P is a statement from which you can deduce $\sim P$. Which, if any, of the following conclusions can you draw:

(a) P is true, (b) P is false, (c) $\sim P$ is true, (d) $\sim P$ is false.

1.20 Give a proof by contradiction of the statement:

P: the sum of the squares of three consecutive integers cannot leave remainder -1 on division by 12.

1.21 Prove by contradiction that the cube of the largest of three consecutive integers cannot be equal to the sum of the cubes of the other two.

1.22 Prove that the polynomial $f(x) = x^4 + 2x^2 + 2x + 1998$ cannot be written as the product of two quadratic polynomials with integer coefficients.

1.23 Prove that if m and n are odd integers, then the equation $x^2 + 2mx + 2n = 0$ has no rational root by deducing that

(a) such a root must be an integer, and

(b) that integer can be neither odd nor even.

1.24 Prove that at a party of at least two people, there are at least two who have the same number of friends at the party.

1.3 Proof by Contraposition

This method may be applied when you wish to prove a compound statement of the form "if P, then Q", where P and Q are themselves statements about the same thing or things. Such a statement is called an **implication**, with premise P and conclusion Q, and it can be put in many different ways:

- Q holds whenever P holds,

- given P, Q is true,

- Q is a consequence of P,

- P is a **sufficient condition** for Q,

- Q is a **necessary condition** for P,

- P **implies** Q, written $P \Rightarrow Q$.

We prefer the last of these and will study it in more detail in the next chapter.

For the moment, note first that the roles of P and Q in such a statement are completely different. To prove that $P \Rightarrow Q$, we would normally start off by assuming that P is true, go through a process of logical deduction, and finish up with the conclusion that Q is true. Now it may be that, having assumed P, the best way to establish Q is by getting a particular contradiction, namely, by assuming $\sim Q$ and deducing $\sim P$. The contradiction "P and $\sim P$" then allows us to conclude that our assumption $\sim Q$ was false, that is, Q is true.

So this proof boils down to showing that $\sim Q \Rightarrow \sim P$; this statement is called the **contrapositive** of $P \Rightarrow Q$. In the next chapter, we shall give a formal proof that the statements $P \Rightarrow Q$ and $\sim Q \Rightarrow \sim P$ are logically equivalent. On the other hand, the implication $Q \Rightarrow P$ is called the **converse** of $P \Rightarrow Q$ and is NOT logically equivalent to it. We sometimes write $P \Leftarrow Q$ instead of $Q \Rightarrow P$, and $P \Leftrightarrow Q$ pronounced "P if and only if Q" instead of "$P \Rightarrow Q$ and $P \Leftarrow Q$".

This prompts a word of warning. When proving an implication $P \Rightarrow Q$, you can NEVER assume that Q is true. It is, after all, what you are trying to prove: if you want to climb a mountain, you don't start at the top. It would save time of course but wouldn't really achieve anything much.

To set up a good example of proof by contraposition, we shall need some useful notation and a familiar definition. Let a and b be integers and consider the condition "b is a multiple of a": $b = ma$ for some integer m. In elementary number theory, it is both convenient and customary to change the emphasis and state this in the form "a divides b", written $a \mid b$. Thus, 1 has only one positive divisor (itself), and every $n > 1$ has at least two (n and 1). Note that $a \mid b$ is *not* a number (like the fraction a/b), but a *statement* about a pair of numbers. Thus, $2 \mid 4$, $2 \mid 2$, $2 \mid 0$ are true, and $2 \mid 1$, $2 \mid 5$, $0 \mid 2$ are false.

Definition 1.3

A positive integer is called **prime** if it has exactly two divisors.

These are, of course, n and 1 (note that 1 is not prime). It is clear that if n is prime, then n is not divisible by any integer a in the range $2 \le a \le \sqrt{n}$. We shall prove by contraposition that this necessary condition for primeness is also sufficient.

Theorem 1.3

If $n \ge 2$ is a positive integer divisible by no integer d in the range $2 \le d \le \sqrt{n}$, then n is prime.

Proof

We have two statements about positive integers $n \ge 2$:

> P n is divisible by no integer d in the range $2 \le d \le \sqrt{n}$,
> Q n is prime.

Proceeding by contraposition, we assume $\sim Q$, that is, n has a divisor a with $1 \ne a \ne n$. Write $n = ab$ with $n \ne b \ne 1$ and consider two cases:

> (i) $a \le b$: then $a^2 \le ab = n$ and $a \le \sqrt{n}$,
> (ii) $a > b$: then $b^2 < ab = n$ and $b \le \sqrt{n}$.

Taking $d = a$ in case (i) and $d = b$ in case (ii), we have a divisor d of n in the range $2 \le d \le \sqrt{n}$. This is the negation of P, and we have proved $\sim Q \Rightarrow \sim P$. Hence, $P \Rightarrow Q$ by contraposition.   □

EXERCISES

Consider the following statements about pairs of positive integers a, b:

$$A \quad : \quad a = b,$$
$$B \quad : \quad a < b,$$
$$C \quad : \quad a \mid b,$$
$$D \quad : \quad a \text{ and } b \text{ have no common prime divisor.}$$

1.25 Write down four more statements A', B', C', D' by interchanging a and b.

1.26 Write down the negations of A, B, C, D.

1.27 Of the 16 implications $P' \Rightarrow \sim Q$, where P and Q range independently over A, B, C, D, exactly five are true. Which ones?

1.28 In each of the remaining 11 cases, write down a specific pair a, b for which P' is true but $\sim Q$ is false.

1.29 Write down the converse of Theorem 1.3. Is it true?

1.30 Jack said: "If there is a hole in this bucket, we won't get down unhurt." A few minutes later, Jill replied: "There must have been a hole in the bucket." Was this deduction correct or not? Explain.

Prove the following three statements by contraposition.

1.31 If n is an integer with n^2 even, then n is even.

1.32 If a and b are real numbers with $a \neq b$, then $\sqrt{ab} \neq (a+b)/2$.

1.33 For positive integers a and b, if $a \neq b$, then $ax^2 + bx + (b-a) = 0$ has no positive integer root.

1.4 Proof by Induction

The methods of proof described in the previous two sections are in common use throughout mathematics. In contrast, the method presented in this section is special: it applies only to statements about positive integers, more precisely, statements of the form "$P(n)$ is true for all positive integers n", where $P(n)$ is a statement involving n. The fact that the method "works" is a consequence of a fundamental property of $\mathbb{N}$ that is best treated as an axiom, although we shall give a "proof" later.

Principle of Mathematical Induction (PMI)

Let $P(n)$ be a statement about positive integers n. Suppose that

(a) $P(1)$ is true, and

(b) $P(n-1) \Rightarrow P(n)$ for all $n > 1$.

Then $P(n)$ is true for all n.

Here, n is called the **inductive variable**,

$$
\begin{array}{ll}
P(1) & \text{the \textbf{inductive base},} \\
P(n-1) & \text{the \textbf{inductive hypothesis} IH, and} \\
P(n-1) \Rightarrow P(n) & \text{the \textbf{inductive step}.}
\end{array}
$$

In many books, the inductive step (b) is written

$$P(n) \Rightarrow P(n+1) \quad \text{for all } n \geq 1, \tag{1.18}$$

which is the same as (b): merely replace n in (b) by $n+1$ throughout (cf. changing the variable of summation in Exercise 1.5). You are free to use either form, whichever best suits the context.

Our first example comprises something we already know, the important special case $a = d = 1$ of Theorem 1.1. It is interesting to compare the proofs in Section 1.1, Exercise 1.11, and the following.

Theorem 1.4

For every positive integer n,

$$P(n): \qquad\qquad \sum_{k=1}^{n} k = \frac{1}{2}n(n+1).$$

Proof

We need to (a) establish the base, (b) carry out the inductive step.

(a) $P(1)$ is true by inspection: both sides are equal to 1 when $n = 1$.

(b) Now suppose that $n > 1$ and assume $P(n-1)$:

IH
$$\sum_{k=1}^{n-1} k = \frac{1}{2}(n-1)n.$$

Then

$$\sum_{k=1}^{n} k = \sum_{k=1}^{n-1} k + n, \qquad \text{by Exercise 1.7,}$$

$$= \frac{1}{2}(n-1)n + n, \qquad \text{by IH,}$$

$$= \frac{1}{2}(n-1+2)n, \qquad \text{by the distributive laws,}$$

$$= \frac{1}{2}n(n+1).$$

We have thus deduced $P(n)$ from $P(n-1)$. By the PMI, $P(n)$ is true for all positive integers n.   □

In the next example, it is not easy to see how to proceed without the PMI.

Theorem 1.5

For every positive integer n,

$$P(n): \qquad \sum_{k=1}^{n} k^2 = \frac{1}{6}n(n+1)(2n+1).$$

Proof

(a) is true by inspection: both sides are equal to 1 when $n = 1$.
    For (b), suppose that $n > 1$ and assume $P(n-1)$:

IH
$$\sum_{k=1}^{n-1} k^2 = \frac{1}{6}(n-1)n(2n-1).$$

Then

$$\sum_{k=1}^{n} k^2 = \sum_{k=1}^{n-1} k^2 + n^2$$

$$= \frac{1}{6}(n-1)n(2n-1) + n^2$$

$$= \frac{1}{6}n(2n^2 - 3n + 1 + 6n)$$

$$= \frac{1}{6}n(2n^2 + 3n + 1)$$

$$= \frac{1}{6}n(n+1)(2n+1).$$

By the PMI, this completes the proof.   □

The inductive method admits a number of useful variants, all based on the PMI. Here are four of them.

Variant 1: Changing the Base

It may be that we wish to prove a statement $P(n)$ for all values of $n \geq n_0$, where n_0 is an integer other than 1. In this case, it is sufficient to prove

(a) $P(n_0)$,   and   (b) $P(n-1) \Rightarrow P(n)$ for all $n > n_0$.

This is no more than the PMI applied to the new statement $Q(n) = P(n + n_0 - 1)$, since

(a) $Q(1) = P(n_0)$,   (b) $n > n_0 \Leftrightarrow n - n_0 + 1 > 1$.

Theorem 1.6

For all $n \geq 5$, $2^n > n^2$.

Proof

Let $P(n)$ be the statement $2^n > n^2$. $P(5) : 32 > 25$.
  Now let $n > 5$ and assume $P(n-1)$.

IH
$$2^{n-1} > (n-1)^2.$$

  Then

$$
\begin{aligned}
2^n &= 2 \cdot 2^{n-1} \\
&> 2(n-1)^2 \\
&= 2n^2 - 4n + 2 \\
&= n^2 + n(n-4) + 2 \\
&> n^2 \text{ as } n > 5.
\end{aligned}
$$

$\square$

Variant 2: Cumulative Induction

In carrying out the inductive step, it is sometimes convenient (even necessary) to assume *all* the statements $P(k)$ for $1 \leq k \leq n-1$ in order to conclude $P(n)$. This is allowed. Given that

(a) $P(1)$ is true, and

(b) $P(1)$ and $P(2)$ and ... and $P(n-1)$ together $\Rightarrow P(n)$ for all $n > 1$,

you can conclude that $P(n)$ is true for all $n \geq 1$. To see this, merely apply the PMI to the new statement $Q(n) = P(1)$ and $P(2)$ and ... and $P(n)$.

Theorem 1.7

Every integer $n \geq 2$ is a product of primes.

Proof

The inductive base $P(2)$ is clear: 2 is a prime. Now let $n > 2$ and assume $P(k)$ for all k in the range $2 \leq k \leq n - 1$.

We need to distinguish two (mutually exclusive and exhaustive) cases: n is prime, n is not prime. In the first case there is nothing to prove, so assume that n is not prime. Then $n = lm$ with $1 < l, m < n$. By $P(l)$, l is a product of primes, and by $P(m)$, so is m. Thus, so also is their product $lm = n$. $\qquad\square$

P.S. You may be worried about the use of the expression "product of primes" being allowed to include the case of a single prime. This is a convention: a product $t_1 t_2 \cdots t_n$ of n terms is just t_1 when $n = 1$. It is also convenient to attach a meaning when $n = 0$: the **empty product** is taken to be 1. Using the notation

$$t_1 t_2 \cdots t_n = \prod_{k=1}^{n} t_k = \pi_n,$$

we take $\pi_0 = 1$, just as in the case of sums,

$$t_1 + t_2 + \cdots + t_n = \sum_{k=1}^{n} t_k = \sigma_n,$$

we take $\sigma_0 = 0$.

Variant 3: Double Induction

Suppose you want to prove a statement of the form $P(m, n)$ for all positive integers m and n. Induction on n would involve:

(a) establishing the inductive base: $P(m, 1)$ for all values of m, and

(b) carrying out the inductive step: for all $n > 1$,

$$P(m, n-1) \Longrightarrow P(m, n) \text{ for all values of } m.$$

Now either or both of these statements are candidates for proof by induction on m. Then again, the roles of m and n in the aforesaid can be reversed. There are thus a number of subvariants of proof by double induction, depending on the *order* in which the pairs (m, n) are taken (this idea will be developed further in Section 4.4). We consider three of these now, spelling out in each case the strategy for proving $P(m, n)$ for all positive integers m, n.

Strategy 1

(a) Prove $P(1, 1)$,

(b) assume $P(m, 1)$

(c) deduce $P(m + 1, 1)$,

(d) assume $P(m, n)$ for all values of m,

(e) deduce $P(m, n + 1)$ for all values of m.

A little thought shows that this is really two separate single inductions: (a) plus (b) and (c) establish (by induction on m) the base for an induction on n (with (b) and (c) forming the inductive step). For an example of this subvariant, see Exercise 1.43.

Strategy 2

(a) Prove $P(1, n)$, for all values of n,

(b) prove $P(m, 1)$ for all values of m,

(c) for $m, n > 1$, assume $P(m - 1, n)$ and $P(m, n - 1)$,

(d) deduce $P(m, n)$.

Here, (a) and/or (b) can be proved by (single) induction, and (c) and (d) may be thought of as a single induction on $m + n$. This method is illustrated in Exercises 1.47–1.49 below.

Strategy 3

(a) Prove $P(m, 1)$, for all values of m,

(b) for $n > 1$, assume $P(m, n - 1)$,

(c) prove $P(1, n)$,

(d) for $m > 1$, assume $P(m - 1, n)$,

(e) prove $P(m, n)$.

This method differs only slightly from Strategy 2, but highlights the fact that the inductive step (in the induction on n) is being carried out by induction (on m). An example of this kind, but with m starting at 0, follows at once.

Theorem 1.8

For every positive integer n, the product of any n consecutive integers is divisible by $n!$.

Proof

Let us write

$$\pi(m, n) = \prod_{k=1}^{n} (m + k), \ n \geq 1, \ m \geq 0$$

for the product of the n consecutive integers starting from $m + 1$.

Notice that if $n \geq 2$, the first $n - 1$ terms comprise $\pi(m, n - 1)$, so that we have

$$\pi(m, n) = (m + n) \, \pi(m, n - 1). \tag{1.19}$$

If also $m \geq 1$, the last $n - 1$ terms of $\pi(m - 1, n)$, also comprise $\pi(m, n - 1)$, so that

$$\pi(m - 1, n) = m\pi(m, n - 1). \tag{1.20}$$

Subtracting (1.20) from (1.19), we get the formula

$$\pi(m, n) - \pi(m - 1, n) = n \, \pi(m, n - 1), \quad m \geq 1, \ n \geq 2, \tag{1.21}$$

which will come in handy later. We are now ready to embark on the proof.

We want to prove the statement

$$P(m, n) : \qquad\qquad n! \mid \pi(m, n) \quad \text{for all } n \geq 1, \ m \geq 0. \tag{1.22}$$

We induct first on $n \geq 1$. The inductive base,

$$P(m, 1) : 1! \mid \prod_{k=1}^{1} (m + k) \quad \text{for all } m \geq 0,$$

merely asserts that 1 is a divisor of $m + 1$.

Now let $n > 1$ and assume the inductive hypothesis

$$(n-1)! \mid \pi(m, n-1) \quad \text{for all } m \geq 0. \tag{1.23}$$

We have to prove that

$$n! \mid \pi(m, n) \quad \text{for all } m \geq 0, \tag{1.24}$$

and to do this, we now induct on $m \geq 0$, keeping $n \geq 1$ fixed. The inductive base at $m = 0$ is again trivial: $n! \mid n!$, as $\pi(0, n) = n!$

Now let $m > 0$ and assume the (second) inductive hypothesis

$$n! \mid \pi(m-1, n). \tag{1.25}$$

Our goal is to prove

$$n! \mid \pi(m, n) \tag{1.26}$$

assuming two inductive hypotheses (1.23) and (1.25). The first of these implies that $n!$ divides the right-hand side of (1.21), and so also the left-hand side:

$$n! \mid (\pi(m, n) - \pi(m-1, n)).$$

Since $n!$ divides the second term by (1.25), it also divides the first, and we have proved (1.26). This completes the induction on m, and we have proved (1.24). This in turn completes the induction on n, as we have proved (1.22), as required.

□

Variant 4: Simultaneous Induction

This final variant applies when $P(n)$ is a compound statement of the form $Q(n)$ and $R(n)$, that is, when we are trying to prove two statements simultaneously. The inductive base simply requires two proofs, of $P(1)$ and of $Q(1)$. For the inductive step,

$$P(n-1) \text{ and } Q(n-1) \Rightarrow P(n) \text{ and } Q(n) \quad \text{for all } n > 1,$$

a number of approaches are possible, depending on the relation between $P(n)$ and $Q(n)$. For example, it is enough to prove

$$P(n-1) \Rightarrow Q(n) \text{ and } Q(n-1) \Rightarrow P(n), \quad \text{both for all } n > 1,$$

or, as in our star example in the next section,

$$\begin{aligned} P(n-1) \text{ and } Q(n-1) &\Rightarrow P(n), \quad \text{and} \\ Q(n-1) \text{ and } P(n) &\Rightarrow Q(n), \quad \text{both for all } n > 1. \end{aligned}$$

Extreme examples of the use of induction are provided by recent important research by the Russian group theorists S. I. Adian and A. Yu. Ol'shanskii and their colleagues on the long-standing Burnside problem. In one case, the inductive base is 667 and there are as many as 92 inductive variables. This is "improved" in the other to around eight inductive variables, but at the expense of up to six simultaneous inductions in many places and an inductive base of around 10^{10}.

EXERCISES

1.34 Prove that for every positive integer n, $\sum_{k=1}^{n} k^3 = \frac{1}{4}n^2(n+1)^2$.

Can you think of any natural reason for the fact that this is equal to $\left(\sum_{k=1}^{n} k\right)^2$?

1.35 Consider the polynomials

$$s_0(x) = x, \; s_1(x) = \tfrac{1}{2}x(x+1),$$
$$s_2(x) = \tfrac{1}{6}x(x+1)(2x+1), s_3(x) = \tfrac{1}{4}x^2(x+1)^2.$$

Observe that for each $m = 1, 2, 3$,

$$s_m(x) = m \int s_{m-1}(x)\, dx + cx,$$

where c is chosen to make $s_m(1) = 1$. Apply this formula with $m = 4$ to get $s_4(x)$, and prove by induction on n that $\sum_{k=1}^{n} k^4 = s_4(n)$.

1.36 What is wrong with the following "proof" that all horses are the same colour? Let $P(n)$ be the statement: in any group of n horses, all are the same colour. This is clearly true when $n = 1$ as any horse is the same colour as itself. Next, take any group of n horses and exclude one. The remaining $n-1$ are the same colour by the IH. Now exclude a different horse, so that the remaining $n-1$ (including the one originally excluded) are the same colour, by the IH again. So all n are the same colour.

1.37 Find a formula for the sum of the first n terms of the geometric progression with first term a and common ratio $r \neq 1$, and prove it by induction using the form (1.18) of the inductive step.

1.38 Find the smallest value n_0 of n for which the statement $2^n > n^3$ is true, then prove it for all $n \geq n_0$ by induction.

1.39 Give a proof by contradiction that among the positive integers there are infinitely many primes.

1.40 Prove that $\sum_{k=1}^{n} k \cdot k! = (n+1)! - 1$ for all $n \geq 1$.

1.41 Prove that for every integer $n \geq 2$

$$\prod_{k=2}^{n} (1 - 1/k^2) = (n+1)/2n.$$

1.42 Put together an overall proof strategy for a double induction in which the inductive base and the inductive step of the first induction are *both* proved by induction.

1.43 Taking the associative law for granted and using the form (1.18) of the inductive step, prove the commutative law

$$m + n = n + m \text{ for all positive integers } m \text{ and } n.$$

1.44 Can you think of a shorter proof of Theorem 1.8?

1.5 Inductive Definition

Suppose you wish to define a quantity $q(n)$ that depends on the positive integer n: such an object is called, in the terms of Chapter 5, a **map with domain** $\mathbb{N}$, or a **sequence**. Suppose further that

(a) $q(1)$ is known, and

(b) $q(n)$ can be expressed in terms of $q(n-1)$ for all $n > 1$.

Then a simple induction on n proves the statement $P(n) : q(n)$ is defined for every positive integer n.

For example, the rules,

(a) $q(1) = 1$,

(b) $q(n) = n \cdot q(n-1)$ for $n \geq 2$

provide the definition of $n!$.

As with inductive proofs, the base need not always be 1. For example, the formulae for $s_0(x)$ and $s_m(x)$ in Exercise 1.35 above comprise an inductive definition of the polynomial $s_m(x)$ for all integers $m \geq 0$. A simpler and more basic example of this kind is as follows.

Definition 1.4

The **powers** of a variable x are defined by

$$\text{(a)} \quad x^0 = 1, \qquad \text{(b)} \quad x^n = x^{n-1}x \quad \text{for all } n \geq 1.$$

It is an inevitable fact that statements about inductively-defined quantities are proved by induction. The so-called *rules of indices* furnish a simple illustration.

Theorem 1.9

For all integers $m, n \geq 0$,

$$\text{(i)} \quad x^m x^n = x^{m+n}, \qquad \text{(ii)} \quad (x^m)^n = x^{mn}.$$

Proof

(i) Induct on n: both sides are equal to x^m when $n = 0$. Let $n \geq 1$ and assume the IH:

$$x^m \cdot x^{n-1} = x^{m+n-1} \text{ for all } m \geq 0.$$

Then

$$\begin{aligned}
x^m \cdot x^n &= x^m \cdot x^{n-1} \cdot x && \text{by definition} \\
&= x^{n+n-1} \cdot x && \text{by the IH} \\
&= x^{m+n-1+1} && \text{by definition} \\
&= x^{m+n} && \text{for all } m \geq 0.
\end{aligned}$$

(ii) Induct on n: both sides are equal to 1 when $n = 0$. Let $n \geq 1$ and assume the IH:

$$(x^m)^{n-1} = x^{m(n-1)} \text{ for all } m \geq 0.$$

Then

$$\begin{aligned}
(x^m)^n &= (x^m)^{n-1} x^m && \text{by definition} \\
&= x^{m(n-1)} x^m && \text{by the IH} \\
&= x^{m(n-1)+m} && \text{by part (i)} \\
&= x^{mn} && \text{for all } m.
\end{aligned}$$

$\square$

As with inductive proofs, more than one inductive variable may be involved in an inductive definition. In this case, a number of strategies are available, perhaps the cleanest being Strategy 2 on page 15. Thus, to define a quantity $q(m, n)$ for all $m, n \geq 0$, it is sufficient to

(a) specify $q(m,0)$ for all $m \geq 0$ and $q(0,n)$ for all $n \geq 0$, and

(b) express $q(m,n)$ in terms of $q(m-1,n)$ and $q(m,n-1)$ for all $m,n \geq 1$.

The following example is very natural, important and (I hope) familiar.

Definition 1.5

Let us define

(a) $b(m,0) = 1$ for all $m \geq 0$ and $b(0,n) = 1$ for all $n \geq 0$, and

(b) $b(m,n) = b(m-1,n) + b(m,n-1)$ for all $m,n \geq 1$.

Then the $b(m,n)$ are called **binomial coefficients**; it is customary to write $b(m,n) = \binom{m+n}{m}$, pronounced "$m+n$ choose m".

A little thought will convince you that this is nothing but a formal definition of Pascal's triangle.

The following properties of the binomial coefficients are all easy exercises. For all $m,n \geq 0$,

(i) $\binom{m+n}{m} = \binom{m+n}{n}$,

(ii) $\binom{m+n}{m} = \frac{(m+n)!}{m!\,n!}$,

(iii) $\binom{m+n}{m}$ is the number of ways of choosing m things from $m+n$.

Property (i) expresses the *symmetry* of $b(m,n)$, (ii) provides a *closed formula*, and (iii) indicates an alternative approach in terms of *combinations*. Arguably the most important manifestation of these numbers is in the famous theorem from which they get their name.

Theorem 1.10 (The Binomial Theorem)

For all $n \geq 0$,

$$(1+x)^n = \sum_{k=0}^{n} \binom{n}{k} x^k. \tag{1.27}$$

Proof

The proof is (necessarily) by induction on n. To get the base, let $n = 0$. Then the left-hand side is equal to 1 (Definition 1.4(a)), and so is the right-hand side:

$$\sum_{k=0}^{0} \binom{0}{k} x^k = \binom{0}{0} x^0 = 1,$$

by Definitions 1.5(a) and 1.4(a).

For the inductive step, let $n \geq 1$ and assume the IH:

$$(1+x)^{n-1} = \sum_{k=0}^{n-1} \binom{n-1}{k} x^k.$$

Then

$$
\begin{aligned}
(1+x)^n &= (1+x)^{n-1}(1+x) \text{ by Definition 1.4(b)} \\
&= \left(\sum_{k=0}^{n-1} \binom{n-1}{k} x^k \right)(1+x) \text{ by the IH} \\
&= \sum_{k=0}^{n-1} \binom{n-1}{k} x^k + \sum_{k=0}^{n-1} \binom{n-1}{k} x^{k+1} \\
&= \sum_{k=0}^{n-1} \binom{n-1}{k} x^k + \sum_{k=1}^{n} \binom{n-1}{k-1} x^k \\
&= \binom{n-1}{0} x^0 + \sum_{k=1}^{n-1} \left(\binom{n-1}{k} + \binom{n-1}{k-1} \right) x^k + \binom{n-1}{n-1} x^n.
\end{aligned}
$$

In the second-last step, we replaced k by $k-1$ throughout the second sum, and in the last step we isolated the first term of the first sum and the last term of the second sum, then combined what was left into a single sum.

Comparing this expression with the right-hand side of (1.27) we have to prove that

(i) $\binom{n-1}{0} x^0 = \binom{n}{0} x^0$,

(ii) $\binom{n-1}{k} + \binom{n-1}{k-1} = \binom{n}{k}$ for $1 \leq k \leq n-1$, and

(iii) $\binom{n-1}{n-1} x^n = \binom{n}{n} x^n$.

Well, both sides of (i) are equal to 1 by Definitions 1.5(a) and 1.4(a), and (iii) also follows from Definition 1.5(a): $\binom{m}{m} = 1$ for all $m \geq 0$. To see (ii), write the formula in Definition 1.5(b) in the form

$$\binom{m+n}{m} = \binom{m+n-1}{m-1} + \binom{m+n-1}{m}.$$

Replacing m by k and n by $n-k$ does the trick.   □

Finally, and again in analogy with inductive proofs, inductive definitions can be cumulative, that is to say, having got the base, $q(1)$ say, $q(n)$ for $n > 1$ can be defined in terms of any or all of the previous values $q(k)$, $1 \leq k \leq n-1$. We round off this section by giving two classical examples.

Definition 1.6

Define a sequence of numbers by setting

$$u_0 = 0, \quad u_1 = 1, \quad u_n = u_{n-2} + u_{n-1} \text{ for } n \geq 2.$$

The u_n, $n \geq 0$, are called the **Fibonacci numbers**. Note that two equations are required for the base because of the form of the inductive part of this definition.

As their name suggests, the Fibonacci numbers were first studied by Leonardo of Pisa at the time of the Renaissance. Since that time, these numbers have been a constant source of recreational mathematics, as well as providing insight into natural phenomena such as phyllotaxis. At the present time, the Fibonacci Society produces its journal, the Fibonacci Quarterly, once every three months. The u_n have an almost limitless number of interesting properties, of which we shall prove just one.

Theorem 1.11

The Fibonacci numbers u_n have the property

$$u_{n-1} u_{n+1} = u_n^2 + (-1)^n \text{ for all } n \geq 1.$$

Proof

When $n = 1$,

$$\text{lhs} = u_0 u_2 = 0(0 + 1) = 0 = 1^2 + (-1)^1 = \text{rhs}$$

and we have the inductive base. Let $n \geq 2$ and assume the IH

$$u_{n-2} u_n = u_{n-1}^2 + (-1)^{n-1}.$$

Then, making free use of the definition,

$$
\begin{aligned}
u_{n-1} u_{n+1} - u_n^2 &= u_{n-1}(u_{n-1} + u_n) - (u_{n-2} + u_{n-1})u_n \\
&= u_{n-1}^2 - u_{n-2} u_n \\
&= (-1)^n, \text{ by the IH.}
\end{aligned}
$$

$\square$

As a second example of definition by cumulative induction, consider the following.

Definition 1.7

Define a sequence of numbers by setting

$$c_1 = 1, \quad c_n = \sum_{k=1}^{n-1} c_k\, c_{n-k} \text{ for } n \geq 2.$$

The c_n, $n \geq 1$, are called the **Catalan numbers**. As mentioned in Exercise 1.12, c_n is just the number of ways of bracketing a sum or product of n terms in a given order. The closed formula for the c_n which follows provides a proof of the otherwise non-obvious fact that $(n+1) \mid \binom{2n}{n}$ for all $n \geq 0$ (cf. Theorem 1.8 above).

Theorem 1.12

The Catalan numbers c_n are given by

$$c_{n+1} = \frac{1}{n+1} \binom{2n}{n} \text{ for all } n \geq 0. \tag{1.28}$$

Proof

The first step in the proof is to solve the following apparently harder problem: find a formula for the number d_n of different bracketings of a product of n terms, say $x_1, x_2, \ldots, x_n$, *in any order*. It turns out that

$$d_1 = 1, \quad d_{n+1} = (4n - 2)d_n, \ n \geq 1. \tag{1.29}$$

The first equation is obvious, and provides the base for an induction on $n \geq 1$. The case $n = 1$ is also obvious: $d_2 = 2$. The next case is more typical, and we spell it out now.

The new term x_3 can be introduced *inside* each of the products $x_1\, x_2$, $x_2\, x_1$ in four different ways. In $x_1\, x_2$ for example, it can precede or follow x_1, or precede or follow x_2, resulting in $(x_3\, x_1)x_2$, $(x_1\, x_3)x_2$, $x_1(x_3\, x_2)$, $x_1(x_2\, x_3)$ respectively. Further, it can occur *outside* $x_1\, x_2$ in two ways: $x_3(x_1\, x_2)$, $(x_1\, x_2)x_3$, giving six new products altogether from $x_1\, x_2$. From $x_2\, x_1$ we get another 6, so that $d_3 = 12$, as required.

The general case is similar. In any particular bracketed product of x_1, x_2, $\ldots, x_n$ there are $n - 1$ multiplications. Inside each of these we can introduce x_{n+1} in four ways as above to give $4(n - 1)$ new products, and there are two ways of putting it outside. Thus, for each of the d_n products of n terms, we get $4n - 2$ with $n + 1$ terms. This establishes (1.29).

Next, to forge a link with the c_n, observe that for each bracketing, the terms can be written in any of $n!$ orders: $d_n = n! \, c_n$, $n \geq 1$. Hence, by (1.29),

$$c_{n+1} = \frac{d_{n+1}}{(n+1)!} = \frac{(4n-2)d_n}{(n+1)!} = \frac{(4n-2)\,n!\,c_n}{(n+1)!} = \frac{(4n-2)}{n+1}\,c_n. \qquad (1.30)$$

This provides the key to the inductive step in proving (1.28), to which we now finally turn.

To get the base, observe that both sides of (1.28) are equal to 1 when $n = 0$. So assume the equation in (1.28) as it stands and compute as follows:

$$
\begin{aligned}
c_{n+2} &= \frac{4n+2}{n+2}\, c_{n+1}, \quad \text{by (1.30)}, \\[2mm]
&= \frac{4n+2}{n+2} \cdot \frac{1}{n+1}\binom{2n}{n}, \quad \text{by the IH}, \\[2mm]
&= \frac{2n+1}{n+2} \cdot \frac{2}{n+1} \cdot \frac{n+1}{n+1}\binom{2n}{n} \\[2mm]
&= \frac{1}{n+2}\binom{2n+2}{n+1},
\end{aligned}
$$

which completes the inductive step. □

EXERCISES

1.45 Given a sequence a_n, $n \geq 1$, formulate inductive definitions of the expressions

$$\sigma_n = \sum_{k=1}^{n} a_n, \qquad \pi_n = \prod_{k=1}^{n} a_n.$$

1.46 Define $x^{-1} = 1/x$ and $x^{-n} = (x^{-1})^n$ for $n \geq 2$. Prove that $(x^m)^{-1} = (x^{-1})^m$ for all integers m. Deduce that the rules of indices in Theorem 1.9 hold for all integers m and n, positive, negative and zero.

1.47 Use double induction (Strategy 2) to prove the symmetry of the binomial coefficients:

$$b(m, n) = b(n, m) \text{ for all } m, n \geq 0,$$

directly from the definition.

1.48 Similarly prove the closed form $b(m, n) = \frac{(m+n)!}{m!\,n!}$ for all $m, n \geq 0$.

1.49 Similarly, prove that $\binom{m+n}{m}$ is the number of ways of choosing m things from $m + n$, for all $m, n \geq 0$.

1.50 Prove that $\binom{2n}{n} = \sum_{k=0}^{n} \binom{n}{k}^2$ for all $n \geq 0$.

1.51 Deduce from Theorem 1.10 the more general form of the binomial theorem

$$(a + b)^n = \sum_{k=0}^{n} \binom{n}{k} a^k b^{n-k},$$

for all $n \geq 0$ and any numbers a and b.

1.52 Prove that, for all $n \geq 0$, the consecutive Fibonacci numbers u_n, u_{n+1} have no common (positive integer) divisor greater than 1. Is this also true for u_n and u_{n+2}?

1.53 Consider the statements

$$P(n) : u_{2n} = u_{n-1}u_n + u_n u_{n+1}, \qquad Q(n) : u_{2n+1} = u_n^2 + u_{n+1}^2$$

about the Fibonacci numbers. Show that

$$P(n) \text{ and } Q(n) \Rightarrow P(n + 1) , \ Q(n) \text{ and } P(n + 1) \Rightarrow Q(n + 1).$$

Use a simultaneous induction to deduce that the statement $P(n)$ and $Q(n)$ is true for all $n \geq 1$.

1.54 Prove that the Fibonacci numbers are given by the formula

$$u_n = (\theta^n - \phi^n)/\sqrt{5} \text{ for all } n \geq 0,$$

where $\theta = (1 + \sqrt{5})/2$ and $\phi = (1 - \sqrt{5})/2$.

1.55 Prove that

$$2 \nmid c_n \Leftrightarrow n = 2^m$$

for some positive integer m, that is, the nth Catalan number c_n is odd when n is power of 2 and even otherwise.

1.6 The Well-ordering Principle

In this section we attempt to justify the PMI by appealing to an assertion that may be intuitively more reasonable. Recall your acceptance, at the age of six, and again about 21 pages ago ((1.17) in Section 1.2), of the idea of a fraction in lowest terms. This is an application of the following fact.

Well-ordering principle (WOP). If A is a property of the positive integers possessed by at least one of them, then there is a least positive integer, l say, with A, that is:

$$\text{(i)} \quad l \text{ has } A, \quad \text{and} \quad \text{(ii)} \quad \text{no } k \text{ with } 1 \leq k < l \text{ has } A.$$

Given a rational number r, let A stand for the property of "being a possible denominator of r", that is, a positive integer b has property A if br is an integer. By definition of rational number, there is at least one such b. The WOP then guarantees that there is a least such, and this is the denominator of r in lowest terms.

Theorem 1.13

The PMI is a consequence of the WOP.

Proof

We shall prove the contrapositive of the implication WOP $\Rightarrow$ PMI. So assume the PMI to be invalid. This means that there is a statement $P(n)$ about positive integers n such that

(a) $P(1)$ is true,

(b) $P(n-1) \Rightarrow P(n)$ for all $n > 1$, but

(c) for some positive integer m, $P(m)$ is false.

Let A be the property that $P(n)$ is false, so that m has property A because of (c). By the WOP, there is a least l with A, so that $P(l)$ is false. So $l \neq 1$ because of (a), whence $l \geq 2$ and $l - 1$ is a positive integer. By the minimality of l, $P(l-1)$ is true, and because of (b), $P(l)$ is true. Contradiction. $\quad\square$

It turns out (Exercise 1.56 below) that the converse of this theorem is also true. Thus the PMI and WOP are logically equivalent, and in some areas of application the WOP is easier to use. Some examples follow, of which the first is the very reasonable assertion that, in the normal process of division, the remainder is less than what you divide by.

Theorem 1.14 (Euclid)

Given positive integers a and b, we can find integers q and r such that

$$a = bq + r \quad \text{and} \quad 0 \leq r < b. \tag{1.31}$$

Proof

If $b \mid a$, then we can take $q = a/b$ and $r = 0$. If $b \nmid a$, let A be the property: being of the form $a - bn$ with n an integer. By taking $n = 0$, we see that a has property A, and by the WOP there is a least positive integer l with A. Put $l = a - bm$ with m an integer. We prove by contradiction that $l < b$. If this is false, we can write $l = b + x$ with $x \geq 0$. Then $x = l - b = a - bm - b = a - b(m+1)$. Since $b \nmid a$, $x \neq 0$ and the last equation then asserts that x has property A. By the minimality of l, $x \geq l = b + x > x$, a contradiction. Thus, $l = a - bm$ and $l < b$. Taking q to be m and r to be l, we get (1.31). $\square$

This theorem forms the basic method for calculating the following important quantity associated with two positive integers.

Definition 1.8

The **highest common factor** of two positive integers a and b is the largest positive integer h such that $h \mid a$ and $h \mid b$; we often write $h = (a, b)$. a and b are called **relatively prime** (or **coprime**) if $(a, b) = 1$.

Some discussion is in order here. Note first that (a, b) always exists, since the number of common divisors of a and b is non-zero and finite: 1 divides a and b, and if c divides a and b, then $1 \leq c \leq \min(a, b)$. Next, the definition applies equally well if a or b (or both) is negative. *One* of them can even be zero: since every integer divides zero, $(a, 0) = |a|$. The highest common factor is thus defined for any integers a, b except the pair $0, 0$.

The key to the problem of calculating (a, b) for a given a and b is provided by (1.31). Since $(a, b) = (b, a)$, we can assume that $a \geq b$. If c divides both b and r, then clearly $c \mid a$. On the other hand, as $r = a - bq$, if d divides both a and b, then $d \mid r$. The common divisors of a and b are thus the same as the common divisors of b and r:

$$c \mid a \text{ and } c \mid b \iff c \mid b \text{ and } c \mid r.$$

It follows that $(a, b) = (b, r)$. The point is that $r < b$, and the problem has been reduced.

If $r = 0$, we are finished, for

$$(a, b) = (b, r) = (b, 0) = b.$$

If not, repeat the process with b, r in place of a, b respectively to get

$$b = rq_1 + r_1, \qquad 0 \le r_1 < r$$

and $(b, r) = (r, r_1)$. If $r_1 = 0$, then $(a, b) = r$ and we are finished. If not, repeat with (r, r_1) in place of (b, r) to get a remainder $r_2 < r_1$, and so on. Proceeding in this way we get a strictly decreasing sequence $b, r, r_1, r_2, \ldots$ of non-negative integers which, after a finite number of steps (at most b), must reach zero: $r_n = 0$ say. Then, from what has been said, $(a, b) = r_{n-1}$. This process for calculating (a, b) is called **Euclid's algorithm**.

An important but rather unexpected property of the hcf is described in the following theorem.

Theorem 1.15

Given positive integers a and b with $(a, b) = h$, we can find integers s and t such that

$$h = sa + tb. \tag{1.32}$$

Proof

Note first that in general, one of s, t will be positive and the other negative. Let A be the property of positive integers that they can be written in the form of the right-hand side of (1.32): a positive integer c has A if $c = ma + nb$ for some integers m and n. Since at least one positive integer, $a + b$ for example, has A, there is a least such, call it $l : l = ma + nb$. We make the claim that $l \mid a$ and prove it by contradiction.

Assume that $l \nmid a$. Then by Theorem 1.14 we can write

$$a = ql + r, \qquad 0 < r < l.$$

Then

$$r = a - ql = a - q(ma + nb) = (1 - qm)a + (-qn)b,$$

so that r has A. The fact that $0 < r < l$ contradicts the minimality of l, and our claim that $l \mid a$ is established. The fact that $l \mid b$ is proved in the same way.

We have shown that l is a common factor of a and b, whence $l \le h$. But h divides the right-hand side of the equation $l = ma + nb$. Thus $h \mid l$, so that $h \le l$. Therefore $h = l = ma + nb$, and we can take $s = m$, $t = n$ to get (1.32).

$\square$

This proof illustrates a weakness in the WOP: it is not constructive. We have shown that s and t satisfying (1.32) *exist*, but no indication is given of how to calculate them. Fortunately, Euclid's algorithm comes to the rescue: s and t can be found by substituting back through the equations that led to $r_{n-1} = (a, b)$.

Example 1.1

Calculate the hcf h of the numbers 89 and 55, and find integers s and t such that

$$h = 89s + 55t.$$

The first steps in the algorithm are as follows:

$$
\begin{aligned}
89 &= 55 \cdot 1 + 34, \\
55 &= 34 \cdot 1 + 21, \\
34 &= 21 \cdot 1 + 13, \\
21 &= 13 \cdot 1 + 8, \\
13 &= 8 \cdot 1 + 5, \\
8 &= 5 \cdot 1 + 3, \\
5 &= 3 \cdot 1 + 2, \\
3 &= 2 \cdot 1 + 1, \\
2 &= 1 \cdot 2 + 0,
\end{aligned}
$$

and so $h = 1$. Working backwards,

$$
\begin{aligned}
1 = 3 - 2 &= 3 - (5 - 3) = 2 \cdot 3 - 5 \\
&= 2(8 - 5) - 5 = 2 \cdot 8 - 3 \cdot 5 \\
&= 2 \cdot 8 - 3(13 - 8) = 5 \cdot 8 - 3 \cdot 13 \\
&= 5(21 - 13) - 3 \cdot 13 = 5 \cdot 21 - 8 \cdot 13 \\
&= 5 \cdot 21 - 8(34 - 21) = 13 \cdot 21 - 8 \cdot 34 \\
&= 13(55 - 34) - 8 \cdot 34 = 13.55 - 21 \cdot 34 \\
&= 13 \cdot 55 - 21(89 - 55) = 34 \cdot 55 - 21 \cdot 89,
\end{aligned}
$$

so that $s = -21$ and $t = 34$.

So the Fibonacci numbers turn up as the canonical worst case of Euclid's algorithm. This is atypical: the algorithm is in general very efficient.

As a consequence of Theorem 1.15 we shall now obtain an important property of prime numbers, which is a necessary preliminary for the last theorem in this chapter.

Theorem 1.16

If n is a prime dividing a product ab of positive integers, then $n \mid a$ or $n \mid b$.

Proof

The proof is by contradiction. So assume that n is a prime with $n \mid ab$, but that n divides nether a nor b. Since n is prime, (n, a) can only be n or 1. Our hypotheses rule out the first possibility, and so n and a are coprime: $(n, a) = 1$. Similarly $(n, b) = 1$. By Theorem 1.15, we can find integers s, t, u, v such that

$$1 = sn + ta, \qquad 1 = un + vb.$$

Then

$$1 = (sn + ta)(un + vb) = sun^2 + (uta + vbs)n + tv\,ab.$$

Since $n \mid ab$, n divides the right-hand side, whence also $n \mid 1$. Contradiction. $\square$

We are now in the happy position of being able to supply a proof of the Fundamental Theorem of Arithmetic.

Theorem 1.17

Every positive integer n can be expressed as a product of primes. Writing

$$n = p_1^{r_1} p_2^{r_2} \cdots p_l^{r_l} \tag{1.33}$$

with $p_1, p_2, \ldots, p_l$ all primes subject to the conditions

$$\text{(i)} \quad p_1 < p_2 < \cdots < p_l, \qquad \text{(ii)} \quad r_1, r_2, \ldots, r_l \text{ all } \geq 1, \tag{1.34}$$

this expression is unique.

Proof

First the existence of such a decomposition is just the assertion of Theorem 1.7 (in the trivial case $n = 1$, the product is empty).

To prove the uniqueness (and this is typical of such proofs) assume that

$$n = q_1^{s_1} q_2^{s_2} \cdots q_m^{s_m} \tag{1.35}$$

is another such decomposition, that is, $q_1, q_2, \ldots, q_m$ are all prime and

$$\text{(i)} \quad q_1 < q_2 < \cdots < q_m, \qquad \text{(ii)} \quad s_1, s_2, \ldots, s_m \text{ all } \geq 1. \tag{1.36}$$

Then we have to prove that the decompositions (1.33) and (1.35) are *identical*, that is,

$$l = m, \quad \text{and} \quad p_k = q_k, \quad r_k = s_k \qquad \text{for all } k, \quad 1 \le k \le l. \tag{1.37}$$

Proceed by (cumulative) induction on n. When $n = 1$, both products must be empty, so that $l = m = 0$ and (1.37) holds. Now let $n \ge 2$ and assume uniqueness for all k with $1 \le k < n$ as the IH.

Let p be the smallest prime dividing n. Then an easy induction (Exercise 1.63 below) based on Theorem 1.16 shows that p divides one of the p_k, $1 \le k \le l$, and condition (i) of (1.34) forces $p = p_1$. Similarly, $p = q_1$, so $p_1 = q_1$.

It follows that

$$\frac{n}{p} = p_1^{r_1-1} p_2^{r_2} \cdots p_l^{r_l} = q_1^{s_1-1} q_2^{s_2} \cdots q_m^{s_m}. \tag{1.38}$$

Since $n/p < n$, we can apply the IH, provided that the analogues of conditions (1.34) and (1.36) hold for (1.38), that is, with r_1, s_1 replaced by $r_1 - 1$, $s_1 - 1$ respectively.

This is so when r_1, s_1 are both at least 2 (the case $p \mid n/p$), and we can deduce (1.37) from the fact that the products in (1.38) are identical, as $r_1 - 1 = s_1 - 1 \Rightarrow r_1 = s_1$. In the other case, $p \nmid n/p$, we have $r_1 = s_1 = 1$, and the analogous conditions are those obtained from (1.34) and (1.36) by removing the terms p_1, r_1, q_1, s_1. The fact that the products in (1.38) are identical again guarantees (1.37), as $l - 1 = m - 1 \Rightarrow l = m$. Thus, (1.37) holds in both cases and the induction is complete. □

EXERCISES

1.56 Give a proof by contradiction of the converse of Theorem 1.13:

$$\text{PMI} \Rightarrow \text{WOP}.$$

1.57 If l is the least positive denominator of a rational number s, prove that the possible denominators are just the positive multiples of l.

1.58 Show that Theorem 1.14 remains true when a is allowed to be negative.

1.59 Let h be the highest common factor of the positive integers a and b. Prove that the common factors of a and b are just the divisors of h.

1.60 Given positive integers a, b their product is a multiple of both. By the WOP, they have a **least common multiple**, often written $[a, b]$. Prove that $(a, b)\, [a, b] = ab$.

1.61 Find the highest common factor of 582 and 285, and express it in terms of these numbers.

1.62 Find the highest common factor of the polynomials $a(x) = x^6 - 1$ and $b(x) = x^3 + x^2 + x + 1$, and express it in terms of them.

1.63 Let p be a prime and $n \geq 2$ an integer. Prove that if p divides a product $b_1 b_2 \cdots b_n$ of positive integers, then p already divides one of the b_k, $1 \leq k \leq n$.

1.64 Let m and n be positive integers and $p_1, p_2, \ldots, p_l$ a list of the primes that divide at least one of them. Write

$$m = \prod_{k=1}^{l} p_k^{r_k}, \qquad n = \prod_{k=1}^{l} p_k^{s_k},$$

where r_k, $s_k \geq 0$ for $1 \leq k \leq l$. Prove that

$$h = \prod_{k=1}^{l} p_k^{t_k}, \qquad t_k = \min(r_k, s_k), \quad 1 \leq k \leq l,$$

is equal to the highest common factor (m, n) of m and n.

1.65 With m, n as in the previous exercise, write down the prime factorization for the least common multiple $[m, n]$ of m and n. Give an alternative solution of Exercise 1.60.

2
Logic

"If you want to play the game, you'd better know the rules."
C. Eastwood,
The Dead Pool

The word **logic** derives from the Greek λογοσ: reasoning, and is defined in the OED as the:

(a) branch of philosophy that deals with reasoning and thinking, especially inference and scientific method;

(b) systematic use of symbolic techniques and mathematical methods to determine the forms of valid deductive argument.

These definitions nicely illustrate the two-way traffic between logic and mathematics. Thus, according to (a), logic underpins mathematics, which is the main reason why this book was written. On the other hand, (b) declares that mathematical ideas, notation and methods can be used to describe and develop the study of logic, and this forms the content of this chapter. The terms "chain of reasoning", "inference", and "valid deductive argument" are all more or less synonymous with **proof**.

2.1 Propositions

The fundamental objects of study in logic are *propositions*. What is a proposition? First and foremost it is a *sentence*: logic is bound up with language. But

What's the time?
Do not lean out of the window!

are sentences and we wish to exclude these. All right, how about *statement?* This is close, but the statements

There will be a sea-battle tomorrow.

The function $f(x) = \cos x$ is commutative.

There are two top-class football teams in Nottingham: Notts County and Notts County Reserves.

are unacceptable for various reasons. The point is that we want our propositions to have a definite *truth-value:* a **proposition** is a statement that is either true or false, but not both.

The (mathematical) statements, denoted by P, Q, IH, etc., in the previous chapter are all of this type. The letters P, Q, IH are being used to represent propositions in much the same way as letters x, a, b are used to represent numbers in ordinary algebra. In both cases, *symbols* (upper or lower case letters) are taken to stand for actual objects (propositions or numbers). Again in both cases, formulae can be put together using these symbols and evaluated by substituting for the symbols objects that they represent. For this reason, the subject we are about to describe and develop is more precisely called *symbolic logic* or *formal logic*. This study is commonly referred to as *propositional calculus*. I'm not too keen on this term: "algebra of propositions" would be more accurate. The word "formal" here emphasizes the fact that we are concentrating on the *form* of symbolic expressions rather than their content. In other words, we are concerned with syntax (grammar) rather than semantics (meaning).

In ordinary algebra, symbols can be combined to form more complicated expressions using certain well-defined operations. Thus, when a, b, c represent numbers, we can attach a meaning to the expression $a(b + c)$, where the operations involved are arithmetical: addition $(+)$ and multiplication (the invisible . between a and $(b+c)$). In just the same way, the symbols in logic that stand for propositions can be combined to form more complicated expressions using logical operations. Thus, when P, Q represent propositions, we can attach a meaning to the expression $\sim P \Rightarrow Q$, where the operations involved are **negation** $(\sim)$ and **implication** $(\Rightarrow)$. In such expressions, P and Q are sometimes called *logical variables*, and the symbols $\sim$ and $\Rightarrow$ *logical constants*. Being mathematicians, however, we prefer the terms proposition and (logical) operation, respectively.

Another two important logical operations are the **connectives** of

conjuction: $P \wedge Q$, "P and Q",

disjunction: $P \vee Q$, "P or Q".

They are defined by their truth-values as follows. $P \wedge Q$ is true when both P, Q are true and is false otherwise. $P \vee Q$ is true when at least one of P, Q is true and is false otherwise. Thus, $\vee$ stands for the "inclusive or": either or both.

Our fifth and final operation is the analogue of equality in algebra, **logical equivalence**: $P \equiv Q$, which is true when P and Q have the same truth-value and false otherwise.

Of the five operation just described, all but $\sim$ are **binary operations**, that is, they combine two propositions to form a third. In contrast, $\sim$ is sometimes referred to as a **unary operation**: it is applied to a *single* proposition.

To conclude this section, we should say a word about the rather special operation $\Rightarrow$. While $P \wedge Q$ and $P \vee Q$ are propositions of much the same kind as P and Q, the expression $P \Rightarrow Q$ carries intuitive overtones of something more: that "Q can be deduced from P". Thus, if P and Q are propositions about numbers, so are $P \wedge Q$ and $P \vee Q$; but whatever the nature of P and Q, $P \Rightarrow Q$ is *always* a proposition about propositions. In the terminology of Chapter 4, $\Rightarrow$ is actually a *relation* between propositions.

There is also some ambiguity here, and there are several different kinds of implication (for example, syntactic implication and semantic implication in Model Theory), and real logicians use different symbols for them. We cheerfully ignore all this and write $P \Rightarrow Q$ in all cases. For our crude purposes, however, it is sufficient to distinguish two kinds, and these differ only in context or emphasis. In the general context of a theorem of the form "if H, then C", the emphasis is on deducing the conclusion C from the hypothesis H, that is, on the proof. In the algebra of propositions dealt with in this chapter, the proposition $P \Rightarrow Q$ is thought of as being defined by its truth-values (see the discussion in the proof of Theorem 1.3 above, Exercise 2.6 below, and the next section).

EXERCISES

2.1 Letting $\triangle$ denote the "exclusive or", write down a definition of $P \triangle Q$ using only the symbols $P, Q, \vee, \wedge, \sim$ and parentheses ( ).

2.2 Boole's fundamental "Laws of Thought" assert that,

$$P \equiv P, \quad P \vee \sim P, \quad \sim (P \wedge \sim P),$$

for all propositions P. Do you believe them?

2.3 By comparing truth-values, prove the commutative laws

$$A \vee B \equiv B \vee A, \quad A \wedge B \equiv B \wedge A$$

for all propositions A, B.

2.4 Similarly, prove the associative laws

$$A \vee (B \vee C) \equiv (A \vee B) \vee C, \quad A \wedge (B \wedge C) \equiv (A \wedge B) \wedge C$$

for all propositions A, B, C.

2.5 Let n be a positive integer and P_k a proposition depending on k for each k, $1 \le k \le n$.

Use truth-values to give sensible definitions of

(a) iterated conjunction: $C = \wedge_{k=1}^{n} P_k$,

(b) iterated disjunction: $D = \vee_{k=1}^{n} P_k$.

2.6 By referring to the proof of Theorem 1.3, give a definition of $P \Rightarrow Q$ in terms of truth-values.

2.7 Define the operation $\Leftrightarrow$ in terms of $\Rightarrow$, $\wedge$ and parentheses. Prove that the propositions $P \Leftrightarrow Q$ and $P \equiv Q$ are logically equivalent.

2.8 Express in logical notation the principles of proof by

(a)   contradiction,   (b)   contraposition,   (c)   induction.

2.9 By considering truth-values, show that there are just four unary operations on propositions. Express their separate effects on a proposition P and think of suitable names for them. How many binary logical operations are there? n-ary?

2.10 For positive integers m and n, let $P(m, n)$ be the proposition $m \mid n$. For how many values of m is $P(m, n)$ true when $n = 16$? For which values of n is $P(m, n)$ true when $m = 16$?

2.11 With $P(m, n)$ as in the previous exercise, translate the following proposition into English:

$$(P(n, ab) \Rightarrow P(n, a) \vee P(n, b)) \Rightarrow (P(m, n) \Rightarrow m = 1 \vee m = n).$$

2.12 Now let $P(m, n)$ be any proposition about pairs m, n of positive integers. What do you make of the following proposition?

$$P(1, 1) \wedge (P(m, n) \Rightarrow (P(m + 1, n) \wedge P(m, n + 1))) \Rightarrow P(m, n).$$

2.2 Truth Tables

Both the *definitions* of logical operations and the *proofs* of laws that hold for them can be performed in a very satisfactory way using truth tables. This method of handling propositions enjoys (at least) three important advantages:

- it is absolutely rigorous and logically correct;

- it presents all the information in a simple visual manner; and

- it can be adapted to the study of sets, as in the next chapter.

Suppose first that we want to define a binary operation, call it $_o$, on propositions. This is done by assigning a truth-value to the proposition $C = P_o Q$ for all possible combinations of truth-values of P and Q. Now any given proposition can take only two possible truth-values, which we denote by 1 (true) and 0 (false). There result four possible input truth-combinations for P and Q, and these correspond to the *rows* of the truth table. The *columns* are headed by the propositions involved, three in this case: $P, Q, P_o Q$.

The *entry*, 0 or 1, in each place is the truth-value of the proposition corresponding to that column for the input truth-combination for that row. The result is a 4×3 "matrix" of zeros and ones. We usually take the P-column to be 1100 and the Q-column to be 1010, whereupon the $P_o Q$-column contains the definition of the operation $_o$. Time for some examples.

Table 2.1. Logical operations defined by truth tables.

P	$\wedge$	Q
1	1	1
1	0	0
0	0	1
0	0	0

P	$\vee$	Q
1	1	1
1	1	0
0	1	1
0	0	0

P	$\Rightarrow$	Q
1	1	1
1	0	0
0	1	1
0	1	0

P	$\equiv$	Q
1	1	1
1	0	0
0	0	1
0	1	0

P	$\triangle$	Q
1	0	1
1	1	0
0	1	1
0	0	0

P	$*$	Q
1	0	1
1	1	0
0	1	1
0	1	0

In Table 2.1 the outer columns give the input truth-values for P and Q, so that the rows correspond to the combinations:

1 P and Q both true,

2 P true and Q false,

3 P false and Q true,

4 P and Q both false.

The middle six columns then *define*:

$P \wedge Q$ as true when P and Q are both true and false otherwise,

$P \vee Q$ as false precisely when P and Q are both false,

$P \Rightarrow Q$ as false only when P is true and Q is false,

$P \equiv Q$ as true only when P and Q have the same truth-value,

$P \vartriangle Q$ as true only when exactly one of P and Q is true,

$P * Q$ as false only when P and Q are both true.

The first four of these are just the definitions of the basic operations $\wedge, \vee, \Rightarrow, \equiv$ given in the previous section. $\vartriangle$ is the usual notation for the "exclusive or", and $*$ is another operation that turns out to have an interesting property (see Exercise 2.23); $P * Q$ means that at least one of P, Q is false.

So much for definitions. Truth tables can also be used to establish *laws*, which assert the logical equivalence of various compound propositions. Examples are the **de Morgan laws**,

$$\sim(P \wedge Q) \equiv \sim P \vee \sim Q, \quad \sim(P \vee Q) \equiv \sim P \wedge \sim Q,$$

commutative laws for $\vee$ and $\wedge$,
associative laws for $\vee$ and $\wedge$,
distributive laws for $\wedge$ over $\vee$ and for $\vee$ over $\wedge$.

Such a law is derived from the definitions in Table 2.1 by constructing a truth table as in the following example, where the columns are numbered only for the purpose of describing the construction.

Table 2.2. A distributive law proved by a truth table.

P	$\vee$	$(Q$	$\wedge$	$R)$	$\equiv$	$(P$	$\vee$	$Q)$	$\wedge$	$(P$	$\vee$	$R)$
1	1	1	1	1	1	1	1	1	1	1	1	1
1	1	1	0	0	1	1	1	1	1	1	1	0
1	1	0	0	1	1	1	1	0	1	1	1	1
1	1	0	0	0	1	1	1	0	1	1	1	0
0	1	1	1	1	1	0	1	1	1	0	1	1
0	0	1	0	0	1	0	1	1	0	0	0	0
0	0	0	0	1	1	0	0	0	0	0	1	1
0	0	0	0	0	1	0	0	0	0	0	0	0
1	2	3	4	5	6	7	8	9	10	11	12	13

The entries in Table 2.2 are made one column at a time in the following way.

First step. Enter the input truth-values of P in columns 1, 7, 11, consistently. Do the same for Q in columns 3 and 9, and for R in columns 5 and 13. ˙

Second step Fill in column 4 using the first definition in Table 2.1 with input truth-values from columns 3 and 5. Do the same for columns 8 and 12 using the second definition in Table 2.1 with input truth-values from the adjacent columns.

Third step Get column 2 from columns 1 and 4 using $\vee$, then column 10 from columns 8 and 12 using $\wedge$.

Fourth step Get column 6 from columns 2 and 10 using the definition of $\equiv$.

The process uses the definitions to work cumulatively from within brackets outwards, starting with the prescribed truth-values of P, Q, R and finishing with the truth-value of the whole expression recorded under $\equiv$ in column 6. Since these eight values are all 1, we conclude that the assertion always holds, and we have proved one of the distributive laws.

In contrast to laws, which assert the equivalence of various compound propositions, we also have several *rules of inference*. The key operation in these is, of course, *implication*, and they are actually used by mathematicians to construct proofs. Some important examples are

$$
\begin{aligned}
\text{modus ponens:} && (P \wedge (P \Rightarrow Q)) &\;\Rightarrow\; Q, \\
\text{modus tollens:} && (\sim Q \wedge (P \Rightarrow Q)) &\;\Rightarrow\; \sim P, \\
\text{proof by contradiction:} && (P \Rightarrow (Q \wedge \sim Q)) &\;\Rightarrow\; \sim P, \\
\text{proof by contraposition:} && (\sim Q \Rightarrow \sim P) &\;\Rightarrow\; (P \Rightarrow Q).
\end{aligned}
$$

Since we made use of the last two in the previous chapter, we prove them now, in Table 2.3.

Table 2.3. Proof by contradiction and proof by contraposition.

$(P$	$\Rightarrow$	$(Q$	$\wedge$	$\sim$	$Q))$	$\Rightarrow$	$\sim$	P
1	0	1	0	0	1	1	0	1
1	0	0	0	1	0	1	0	1
0	1	1	0	0	1	1	1	0
0	1	0	0	1	0	1	1	0

$(\sim$	Q	$\Rightarrow$	$\sim$	$P)$	$\Rightarrow$	$(P$	$\Rightarrow$	$Q)$
0	1	1	0	1	1	1	1	1
1	0	0	0	1	1	1	0	0
0	1	1	1	0	1	0	1	1
1	0	1	1	0	1	0	1	0

EXERCISES

2.13 Draw up truth tables defining the four unary operations $P, \sim P, T, F$.

Use truth tables to prove the following seven logical laws.

2.14 Boole's laws of thought.

2.15 The de Morgan laws.

2.16 The distributivity of $\wedge$ over $\vee$.

2.17 The law of permutation,

$$(P \Rightarrow (Q \Rightarrow R)) \equiv (Q \Rightarrow (P \Rightarrow R)).$$

Similarly establish the following five rules of inference.

2.18 Modus ponens and modus tollens.

2.19 The law of syllogism,

$$(P \Rightarrow Q) \Rightarrow ((Q \Rightarrow R) \Rightarrow (P \Rightarrow R)).$$

2.20 The law of importation,

$$(P \Rightarrow (Q \Rightarrow R)) \Rightarrow ((P \wedge Q) \Rightarrow R).$$

2.21 The law of exportation,

$$((P \wedge Q) \Rightarrow R) \Rightarrow (P \Rightarrow (Q \Rightarrow R)).$$

2.22 Show that each of the 16 binary operations possible on propositions can be expressed in terms of $\vee, \wedge, \sim$.

2.23 Show that each of $\sim, \wedge, \vee$ can be expressed in terms of the single binary operation $*$ defined in Table 2.1.

2.3 Syllogisms

The propositional calculus studied the previous two sections is conceptually simpler, though historically later, than another fundamental branch of formal logic, the predicate calculus, which is also known as the logic of quantifiers and forms the foundation for much of the material in this book. This subject had its real beginning in the syllogistic of Aristotle around 350 BC, and this section is devoted to an essay on that ancient topic.

In predicate calculus, the basic objects are terms, which occur as parts, the subject and predicate, of a proposition. **Terms** are "things", like parrots or numbers, or (in the case of predicates) "attributes", like mortal or prime. (The distinction between a thing and an attribute is little more than a linguistic convention: an attribute becomes a thing if we put the word "thing" after it.)

A **syllogism** is an inference of the form $P \wedge Q \Rightarrow R$, where the **premises** P, Q and the **conclusion** R are all **categorical propositions**, that is, each is one of the following four types:

A: every s is (a) p,

E: no s is (a) p,

I: some s is (a) p,

O: some s is not (a) p,

In each case, the **subject** s and **predicate** p are terms, "some" means "at least one", and three different terms appear twice each in different propositions. This is illustrated by the following classical example.

P no god is mortal,

Q every man is mortal,

R no man is a god,

where Q is of type A, and P, R are of type E.

In order to enumerate the possible syllogistic forms, we need some nomenclature. First, of the three terms involved, the predicate of the conclusion is called the **major term**, the subject of the conclusion is the **minor term**, and the third term is the **middle term**. Then the premise containing the major term is the **major premise** and that containing the minor term is the **minor premise**. In the above example, the major, minor and middle terms are thus god, man, mortal respectively. It is customary, as in this example, to *write the major premise first.*

With this in mind, one more ingredient is needed to complete the description of a given syllogistic form: the distribution of the middle term between the two premises. This can be done in four ways, corresponding to the four **figures** of the syllogism. Letting c denote the major term and a the minor, these figures are as in Table 2.4, where b denotes the middle term.

Table 2.4. The four figures of the syllogism.

1	2	3	4
b c a b	c b a b	b c b a	c b b a

The subject is written first in each case, and so (obviously) is the major premise. In the above example the middle term (mortal) appears as the predicate in both premises, so this is an example of the second figure.

For a more mathematical example, consider the syllogism:

major premise	:	no square is prime	(type E)
minor premise	:	some squares are odd	(type I)
conclusion	:	some odd numbers are not prime	(type O).

The middle term $b = $ "square" appears as the subject in both premises, so this is an example of figure 3.

Our enumeration of syllogistic forms is now complete: there are four types (A, E, I, O) possible for each of the three propositions involved and four figures (1, 2, 3, 4) for each such combination.. So there are just 256 altogether. Of these, it was shown by Aristotle and his pupil Theophrastus that exactly 15 represent valid deductions. Those of a given figure can be represented by words containing three vowels corresponding in order to the types of the three propositions. This was actually done in mediæval times, and the resulting mnemonics appear in Table 2.5. A quick check shows that our two examples above are of the form CESARE and FERISON respectively, and so both are valid.

Table 2.5. The 15 valid syllogistic forms.

1	2	3	4
BARBARA	CESARE	DISAMIS	CAMENES
CELARENT	CAMESTRES	DATISI	DIMARIS
DARILI	FESTINO	BOCARDO	FRESISON
FERIO	BAROCO	FERISON	

We turn attention to the problem of establishing the validity or otherwise of a given syllogistic form. The method described below, while of some historical

interest, is not entirely satisfactory. A better method will emerge in the course
of the next chapter.

The method involves the use of a **Lewis Carroll diagram**. This takes
the form of a square U as shown in Fig. 2.1, with the upper half, left half
middle square representing the major, minor, middle terms respectively, and
the complementary parts representing their negations in each case. The square
is thus divided up into eight sections. Taking CESARE in as in our first example
above, the major premise asserts that "no c is b", that is, the upper half of the
middle square is empty, and we record this by putting zeros (0) in the two
corresponding sections of U. The minor premise asserts that "every a is b", or
"no a is not b", and two more sections can be marked with zero for empty.
The conclusion "no a is a c" will hold if the upper left quadrant of U is empty,
and indeed it is. So the conclusion holds and CESARE is valid in the second
figure.

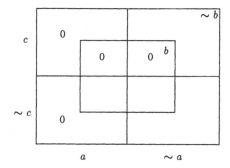

Fig. 2.1. Lewis Carroll diagram for CESARE.

Propositions of type A and E correspond to the assertion that two sections
of the diagram are empty. Types I and O correspond to the assertion that at
least one of two sections is non-empty, but we cannot say which. To indicate
this, we put a one ($|$ or possibly $-$) on the common boundary of the sections
concerned. To illustrate this, consider the following example:

P some parrots can talk,

Q all talkers are human beings,

R some human beings are not parrots,

which is of the type TIRANO in the fourth figure. We annotate Fig. 2.2 with p
for the major term, h for the minor term and t for the middle term. The $-$
between left and right halves of the upper half of the middle square indicates

that one of these two regions is not empty. The minor premise gives rise to the two zeros.

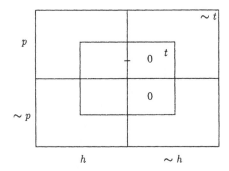

Fig. 2.2. Invalidity of TIRANO in the fourth figure.

We want to deduce that the lower left quadrant of the diagram is non-empty. But the markings corresponding to the premises do not touch this part of the diagram. Hence, we cannot make the conclusion and TIRANO in the fourth figure is invalid.

Incidentally, we *can* deduce from the diagram that the top left quarter of the middle square is non-empty, that is, "some h are p", which is of type I. Hence, DIMARIS is valid in the fourth figure.

While the previous proof (of validity) is quite convincing, this proof (of invalidity) seems somehow less satisfactory. While it can be justified, you might prefer an alternative method, such as giving a *counter-example*. To show that a universal statement, like "I and A *always* imply O in the fourth figure", is false, it is sufficient to find *one example* where it fails. Here is a case of TIRANO in the fourth figure where P and Q are true and R is false:

P some integers are multiples of 4

Q all integer multiples of 4 are even

R some even numbers are not integers.

EXERCISES

Use Lewis Carroll diagrams to prove the validity of the following three syllogistic forms.

2.24 CELARENT in the first figure.

2.25 FERISON in the third figure.

2.26 DIMARIS in the fourth figure.

Give counter-examples to prove the invalidity of ORLANDO in each of the following three figures.

2.27 The first.

2.28 The second.

2.29 The fourth.

2.30 It can be contended that, as $A \Rightarrow I$ and $E \Rightarrow 0$, nine more valid forms can be added to the list in Table 2.5: BARBARI and CELARONT in the first figure, CESARO and CAMESTROS in the second, DARAPTI and FELAPTON in the third, and BRAMANTIP, CAMENOS and FESAPO in the fourth. By experimenting with these, try to put your finger on the root of this contention and so resolve it.

2.31 Lewis Carroll diagrams can be used as an alternative to truth tables for proving laws and rules of inference of the type discussed in Section 2 above. Thus, if a, b, c are replaced by A, B, C respectively in Fig. 2.1, the eight sections of the diagram correspond in the obvious way to the eight input combinations of truth-values for any expression X involving propositions A, B, C. Placing x in just those sections for which X is true then gives a complete description of the truth-values of X. Apply this in the case

$$P = B \mathbin{\triangle} C, \quad X = A \mathbin{\triangle} P, \quad Q = A \mathbin{\triangle} B, \quad Y = Q \mathbin{\triangle} C$$

to prove the associative law for $\triangle$ (the exclusive or).

2.32 Using the fact that $(P \Rightarrow Q) \equiv (\sim P \vee Q)$, use a Lewis Carroll diagram to prove the law of syllogism

$$(A \Rightarrow B) \Rightarrow ((B \Rightarrow C) \Rightarrow (A \Rightarrow C)).$$

2.33 Here is a four-term syllogism of type IMPERATOR:

> some beasts are cats
>
> no dogs are cats
>
> all beasts are animals
>
> $\therefore$ some animals are not dogs.

Is it valid or not?

2.4 Quantifiers

Recall Exercise 2.8(c), in which the PMI was put into the notation of propositional calculus as follows:

$$(P(1) \wedge (P(n) \Rightarrow P(n+1))) \Rightarrow P(n),$$

where n is a positive integer. The propositions $P(n) \Rightarrow P(n+1)$ on the left and $P(n)$ on the right need to be made more precise by specifying the range of values of n we have in mind. In both cases, that range is the whole of $\mathbb{N}$, as in the statement of the PMI in Section 1.4. This is expressed succinctly using the **universal quantifier** $\forall$, pronounced "for every" (for all, for each, for any):

$$(P(1) \wedge (\forall n\ P(n) \Rightarrow P(n+1))) \Rightarrow \forall n\ P(n).$$

Similarly, the statement of Theorem 1.4 is just

$$\forall n \sum_{k=1}^{n} k = \frac{1}{2} n\,(n+1), \qquad (2.1)$$

and that of Exercise 2.12 takes the form

$$P(1,1) \wedge (\forall m\ \forall n\ P(m,n) \Rightarrow (P(m,+1,n) \wedge P(m,n+1))) \Rightarrow \forall m\ \forall n\ P(m,n). \qquad (2.2)$$

The double quantifier $\forall m\ \forall n$ is sometimes further abbreviated to $\forall m, n$: "for all m and n" (ranging independently over $\mathbb{N}$). The assumption that m and n are integers is implicit; we will see in the next section how to incorporate this information explicitly into the formula.

Most decent theorems are of a general nature, that is, they begin with a universal quantifier: $\forall x\ P(x)$, so that $P(x)$ is true for all values of the variable x in a specified range. To prove such a theorem, we must show that $P(x)$ is true for every value of x. On the other hand, if just one value of x can be found for which $P(x)$ is false, the proposition $\forall x\ P(x)$ will be false, and we will have proved its negation. Thus,

$$(\text{there is an } x \text{ with } P(x) \text{ false }) \Rightarrow \sim(\forall x\ P(x)).$$

The left-hand part is expressed succinctly using the **existential quantifier** $\exists$, pronounced "there is" (there exists, for some, for at least one):

$$\exists x \sim P(x) \Rightarrow \sim(\forall x\ P(x)). \qquad (2.3)$$

When proving universal propositions false in this way, the x for which $P(x)$ is false is often called a **counter-example**. We now have a complete list of quantifiers: $\forall$ and $\exists$.

To continue with the theme of negating quantified propositions, first observe that the implication (2.3) is actually an *equivalence*:

$$\exists x \sim P(x) \equiv \sim(\forall x \ P(x)), \qquad (2.4)$$

for if $P(x)$ is not true for all x, it must be false for some x. Next, consider the following two laws of propositional calculus:

$$\sim(\sim P) \equiv P, \quad (P \equiv Q) \Rightarrow (\sim P \equiv \sim Q). \qquad (2.5)$$

Starting with (2.4), proceed in four steps: apply the second law, apply the first law, replace $P(x)$ by $\sim P(x)$, apply the first law again. We get the following sequence of equivalences:

$$
\begin{aligned}
\sim(\exists x \sim P(x)) &\equiv& \sim(\sim \forall x \ P(x)), \\
\sim(\exists x \sim P(x)) &\equiv& \forall x \ P(x), \\
\sim(\exists x \sim(\sim P(x))) &\equiv& \forall x \sim P(x), \\
\sim(\exists x \ P(x)) &\equiv& \forall x \sim P(x).
\end{aligned}
\qquad (2.6)
$$

Quantified propositions are thus negated according to the following rules.

Theorem 2.1

$$
\begin{aligned}
\sim(\forall x \ P(x)) &\equiv& \exists x \sim P(x) \\
\sim(\exists x \ P(x)) &\equiv& \forall x \sim P(x).
\end{aligned}
$$

$\square$

So much for singly quantified propositions and their negations. Passing to doubly quantified propositions, we have already had an example: $\forall m, n \ P(m, n)$ in formula (2.2). Reading this as "for every pair (m, n) of positive integers $P(m, n)$ is true", we can regard it as singly quantified. The same applies to its negative, $\exists m, n \sim P(m, n)$: there is a pair (m, n) of positive integers for which $P(m, n)$ is false.

But when two different quantifiers appear, the situation is more complicated. For example, consider the proposition

$$P \qquad\qquad \forall n \ \exists m \qquad m > n, \qquad (2.7)$$

where m and n are positive integers, which asserts that there is no biggest positive integer. Since for any n we can take $m = n+1$ and get a true statement

$(n+1 > n)$, P is true. Another example is obtained by changing the order of quantification:

$$Q \qquad\qquad \exists m\ \forall n \quad m > n. \qquad\qquad (2.8)$$

This asserts something quite different, that there is a positive integer bigger than every positive integer, so Q is false: taking $n = m$ gives a counter-example to the proposition $\forall n\ m > n$. Thus, $\sim(P \equiv Q)$, and we have proved (by counter-example) that quantifiers do not satisfy the commutative law.

Our final result in this chapter concerns the negation of multiply quantified propositions. To negate the proposition in (2.7), proceed as follows:

$$
\begin{aligned}
\sim P &= \sim(\forall n\ \exists m\ m > n) \\
&= \exists n \sim(\exists m\ m > n) \\
&= \exists n\ \forall m \sim (m > n) \\
&= \exists n\ \forall m\ m \leq n.
\end{aligned}
$$

This clearly asserts the existence of a biggest positive integer, and is of course false. Similarly, from (2.8) we get the true statement

$$\sim Q = \forall m\ \exists n\ m \leq n.$$

These formulae suggest the general rule for negating multiply quantified propositions. Its proof, which is by induction on n, is left as an exercise.

Theorem 2.2

Let $P(x_1, \ldots, x_n)$ be a proposition depending on n variables x_k, $1 \leq k \leq n$, and let Q_k, $1 \leq k \leq n$, be n quantifiers. Denoting a succession of quantifications by the product symbol $\prod$ and letting $\overline{\forall} = \exists$, $\overline{\exists} = \forall$,

$$\sim \prod_{k=1}^{n} Q_k\, x_k\ P(x_1, \ldots, x_n) \equiv \prod_{k=1}^{n} \overline{Q}_k\, x_k\ \sim P(x_1, \ldots, x_n). \qquad (2.9)$$

□

One last piece of nomenclature. It may be that in a proposition depending on several variables, not all the variables are quantified. Such variables are called **free**, as opposed to the quantified ones, which are called **bound**. The resulting proposition then depends on the free variables. For example, the proposition $\forall x\ \exists y\ P(x, y, z)$ has two bound variables, x and y, and one free variable, z; it is thus the form of $Q(z)$.

EXERCISES

2.34 Comment on the validity of the deduction

$$\forall x\ P(x) \Rightarrow \exists x\ P(x).$$

(a) when x ranges over the positive integers, and

(b) otherwise.

2.35 Prove the laws appearing in (2.5).

2.36 Prove that
$$\exists x\ \forall y\ P(x,y) \Rightarrow \forall y\ \exists x\ P(x,y).$$

2.37 Show that the converse of the assertion in the previous exercise is false by finding a counter-example, that is, a proposition $P(x,y)$ for which it fails.

2.38 Give a proof of Theorem 2.2.

2.39 Translate the following proposition, about positive integers a, b, c, d, into English:

$$\forall a\ \forall b\ \exists c\ (a \mid c) \wedge (b \mid c) \wedge ((a \mid d \wedge b \mid d) \Rightarrow c \leq d).$$

2.40 Write down the negation of the proposition in the previous exercise.

Quantifiers can also be used to make definitions. Say what kind of numbers x are defined in the following examples, where m and n range over the positive integers.

2.41 $\forall x\ \exists n\ 2n = x.$

2.42 $\forall x\ \exists m, n\ mx = n.$

2.43 $\forall x\ \exists n\ x + n = 0.$

3

Sets

"Of all the words in the English Language
the one with the greatest number of meanings
according to the OED is the word 'set'."
D. L. Johnson
Elements of Logic via Numbers and Sets

And that's not counting proper nouns, such as the name of an evil ancient Egyptian deity, or foreign words, like the Korean word for the number three. By way of contrast, the number of definitions of the word "set" in this book is none. The reason for this is that in modern mathematics the notion of set is fundamental. For example, numbers can be defined in terms of sets, and this will occupy us towards the end of the book, with the object of gaining some kind of comprehension of the infinite.

It is easy to find synonyms for the word "set": family, collection, ensemble, aggregate and sometimes system or class, but never group. It is also easy to give examples where any number of identifiable objects may be thought of as compromising a set. Thus, a number of:

people, pieces of cutlery, golden daffodils, ideas, letters, ravens, theories

may collectively be referred to as a(n):

congregation, canteen, host, theory, alphabet, unkindness, philosophy

respectively. In every case, a number of things are put together and *regarded as a single entity*.

3.1 Introduction

The individual objects that make up a set are called its **elements, members** or (sometimes) **points**. We usually denote sets by capital letters and their members by lower-case letters. Membership is expressed by the symbol $\in$: $a \in A$, pronounced "a is an element of A" or "a belongs to A", and non-membership by $\notin$. The most obvious way to describe a particular set is to list its members, separated by commas and enclosed in braces. For example, the equation

$$A = \{1, 2, 3, 4\} \tag{3.1}$$

asserts that the set A consists of the first four positive integers. Then $3 \in A$ and $5 \notin A$.

If S is any set all of whose elements belong to another set T, we say that S is a **subset** of T, or S is **contained** in T, and write $S \subseteq T$. Put into mathematical notation, this definition takes the form

$$\forall x \in S \quad x \in T,$$

or alternatively

$$x \in S \Rightarrow x \in T.$$

It is by checking this implication that we normally verify the assertion that $S \subseteq T$. Two sets are **equal** when they consist of the same elements, and this happens if and only if each is contained in the other:

$$S = T \equiv S \subseteq T \wedge T \subseteq S.$$

Some related symbols are defined as follows:

$$\begin{aligned} S \supseteq T &\equiv T \subseteq S, \\ S \subset T &\equiv S \subseteq T \wedge S \neq T, \\ S \nsubseteq T &\equiv \sim(S \subseteq T). \end{aligned}$$

The brace notation can be extended to describe subsets. If S is a set and P is some property that the elements of S may or may not have, the expression

$$T = \{x \in S \mid x \text{ has } P\}$$

defines T to be the "set of all elements S that have property P". Thus, if A is as above (3.1) and $\mathbb{N}$, as usual, denotes the set of positive integers, then

$$A = \{n \in \mathbb{N} \mid n \leq 4\},$$

and

$$\{a \in A \mid a \text{ is even}\} = \{2, 4\}.$$

This notation can be used to make an important definition. For any subset A of a given set S, we define the **complement** A' of A in S to be the set of those x in S that do not belong to A:

$$A' = \{x \in S \mid x \notin A\},$$

pronounced "A dashed".

Example 3.1

As an illustration of the power of set theory, we will now show how even the rudimentary ideas and notation described so far are sufficient for a complete treatment of the syllogistic.

We begin by translating the four types of categorical proposition into relations between sets. This is done in two stages, of which the first is merely to abbreviate them using quantifiers and the symbol $\in$. Thus, replacing terms (like parrot) by the corresponding sets (like {parrots}, the set of all parrots), type A takes the form $\forall x \in S \; x \in P$, where S is the subject-set and P the predicate-set. Types E, I, O are similarly abbreviated, and the first stage of the translation is complete:

$$
\begin{aligned}
A &= \forall x \in S \; x \in P, \\
E &= \forall x \in S \; x \notin P, \\
I &= \exists x \in S \; x \in P, \\
O &= \exists x \in S \; x \notin P.
\end{aligned}
$$

For the second stage, notice first that A is just an abbreviation of our definition of the notion of containment: $S \subseteq P$. Next, by our rules for negating quantified propositions (Theorem 2.1), $O = {\sim}A \equiv {\sim}(S \subseteq P) = S \nsubseteq P$. Turning to E, our definition of the complement of a set can be restated in the form: $x \in P' \Leftrightarrow x \notin P$. Finally, $I = {\sim}E \equiv {\sim}(S \subseteq P') = S \nsubseteq P'$. We summarize this into a theorem.

Theorem 3.1

The four types of categorical proposition are expressed in the notation of set theory as follows:

$$
\begin{aligned}
A &= \forall x \in S \; x \in P \equiv S \subseteq P, \\
E &= \forall x \in S \; x \notin P \equiv S \subseteq P' \\
I &= \exists x \in S \; x \in P \equiv S \nsubseteq P' \\
O &= \exists x \in S \; x \notin P \equiv S \nsubseteq P.
\end{aligned}
$$

$\square$

Now a typical syllogism takes the form $P \wedge Q \Rightarrow R$, where each of the propositions P, Q, R is one of these four types, and the three terms involved (which are now sets) are distributed in one of four ways (the four figures). Then the translation of BARBARA (Fig. 1) looks like this:

$$B \subseteq C \wedge A \subseteq B \Rightarrow A \subseteq C, \tag{3.2}$$

which is almost obvious (see Exercise 3.7).

Validity of syllogistic forms is thus easily established using set theory. Showing invalidity in this way is also a simple matter. To illustrate the method, take the form PORSENA in the third figure:

$$B \not\subseteq C \wedge B \subseteq A' \Rightarrow A \subseteq C?$$

To invalidate this, just one counter-example will suffice, that is, three sets A, B, C for which

$$B \not\subseteq C \wedge B \subseteq A' \wedge A \not\subseteq C.$$

In other words, we seek three sets A, B, C such that A and B have no element in common and each has an element not in C. Such sets can be found in the set $S = \{1, 2, 3\}$, namely, $A = \{1\}$, $B = \{2\}$, $C = \{3\}$, and form the required counter-example.

The derivation of the 15 valid syllogistic forms (and the invalidation of the other 241) would make a nice project, and the set-theoretic method just described recommends itself as the best way to set about it.

We conclude this section by forging the first links between the notions of set and number. The number of elements in a set S is called its **cardinality** (or order or power) and denoted by $|S|$ (or $\#S$). A set is finite if its cardinality is finite and infinite otherwise. A set with just one element is called a **singleton**. It is convenient to operate with a set of cardinality 0, called the **empty set** and written $\emptyset$ (Scandinavian letter, not Greek "phi"). It makes sense to say that $\emptyset \subseteq S$ for any set S.

Letting $\mathbb{Z}$, as usual, denote the set of all integers, define

$$\mathbb{Z}(n) = \{x \in \mathbb{Z} \mid 1 \le x \le n\}, \tag{3.3}$$

so that A in (3.1) is just another name for $\mathbb{Z}(4)$. Now for any $n \in \mathbb{Z}$, $n \ge 0$, $|\mathbb{Z}(n)| = n$. Thus, the cardinalities possible for finite sets are precisely the non-negative integers. In Chapter 6 we shall develop a theory of infinite sets, like $\mathbb{N}, \mathbb{Z}, \mathbb{Q}$ (the rational numbers) and $\mathbb{R}$ (the real numbers). In this theory, sets with the same cardinality as $|\mathbb{N}|$ will be called **countably infinite** (or denumerable); the common value of their cardinality is written $\aleph_0$, pronounced "aleph-null". While $\mathbb{N}$ (of course), $\mathbb{Z}$ and $\mathbb{Q}$ are all countably infinite, it turns out that $\mathbb{R}$ is not. Writing $|\mathbb{R}| = c$, the **cardinal of the continuum**, we will prove that $\aleph_0 < c$. So some infinities are bigger than others.

EXERCISES

3.1 Set-theoretic notation can be used to further improve propositions
depending on quantified variables by making precise the range of
these variables. Thus, the formula in (2.1) of Section 2.4 achieves
the definitive form:

$$\forall n \in \mathbb{N} \; \sum_{k=1}^{n} k = \frac{1}{2} n(n+1).$$

Write formulae (2.2), (2.7), (2.8) of Section 2.4 in this form.

3.2 Describe in words the following subsets of the Cartesian plane $\mathbb{R}^2$:

(i) $\{(x,y) \mid x = 0\}$,

(ii) $\{(x,y) \mid y > 0\}$,

(iii) $\{(x,y) \mid x^2 + y^2 = 1\}$,

(iv) $\{(x,y) \mid x \in \mathbb{Z}, \, y \in \mathbb{Z}\}$.

3.3 Of what general geometric type are the following configurations of
points in $\mathbb{R}^2$?

(i) $\{(x,y) \mid x^2 + y^2 \leq 1\}$,

(ii) $\{(x,y) \mid y = ax^2 + bx + c\}$, where $a, b, c \in \mathbb{R}$,

(iii) $\{(x,y) \mid xy = 1\}$,

(iv) $\{(x,y) \mid |x| + |y| \leq 1\}$,

where $|x| = x$ if $x \geq 0$ and $|x| = -x$ if $x < 0$.

3.4 What general name would you give to the following subsets of $\mathbb{R}^3$?

(i) $\{(x,y,z) \mid x^2 + y^2 + z^2 = 1\}$,

(ii) $\{(x,y,z) \mid x + y + z = 1\}$,

(iii) $\{(x,y,z) \mid 0 \leq x \leq 1, \, 0 \leq y \leq 1, \, 0 \leq z \leq 1\}$,

(iv) $\{(x,y,z) \mid x^2 + y^2 = 1\}$.

3.5 Let $'$ denote complementation within a given set S. Prove that for
any $A \subseteq S$, $(A')' = A$.

3.6 Let $'$ denote complementation within a given set S. For $A, B \subseteq S$,
prove that

$$A \subseteq B \Leftrightarrow B' \subseteq A'.$$

3.7 Prove the assertion of (3.2):

$$B \subseteq C \wedge A \subseteq B \Rightarrow A \subseteq C.$$

3.8 Translate the syllogism FRESISON in the fourth figure into a statement about sets, then prove this statement.

3.9 Give a counter-example to prove the invalidity of CALIBAN in the second figure.

3.10 Let $S_{m,n} = \{k \in \mathbb{N} \mid m \leq k \leq n\}$, where $m, n \in \mathbb{N}$. Write down necessary and sufficient conditions for $S_{m,n} \subseteq S_{p,q}$.

3.11 With $S_{m,n}$ as in the previous exercise, what is $|S_{m,n}|$?

3.12 Suggest a scheme whereby the integers $\mathbb{Z}$ can be counted, that is, number the elements of $\mathbb{Z}$ with positive integers $n \in \mathbb{N}$ in such a way that

(i) every $z \in \mathbb{Z}$ is given a number,

(ii) no two different $z_1, z_2 \in \mathbb{Z}$ get the same number,

(iii) every number $n \in \mathbb{N}$ is used.

3.13 Suggest a partial numbering for the positive rationals $\mathbb{Q}^+$, that is, a numbering satisfying (i) and (ii) of the previous exercise.

3.14 Let S be a set with n elements, where n is a non-negative integer. Use the binomial theorem with $x = 1$ to find a formula for the total number of subsets of S.

3.2 Operations

We saw in Section 2.2 how new propositions can be obtained from given ones using logical operations: $P \wedge Q$, $P \vee Q$, $\sim P$, and so on. In a very similar way, set-theoretical operations are rules for getting "new sets from old". This analogy works to advantage in two directions. First, most of the ideas developed in this section and the next will look natural and familiar in the light of their logical counterparts in the previous chapter. Second, the translation into set theory both reinforces and even clarifies our understanding of formal logic.

We begin with four important binary operations. Each of these can be applied to any pair A, B of subsets of a previously specified set S to produce a third subset.

The **intersection** of A and B is the set of elements of S that belong to A and B:

$$A \cap B = \{x \in S \mid x \in A \land x \in B\},$$

pronounced "A intersect B".

The **union** of A and B is the set of elements of S that belong to A or B:

$$A \cup B = \{x \in S \mid x \in A \lor x \in B\},$$

pronounced "A union B".

The **difference** of A and B is the set of elements of S that belong to A but not B:

$$A \setminus B = \{x \in S \mid x \in A \land x \notin B\},$$

pronounced "A minus B".

The **symmetric difference** of A and B is the set of elements of S that belong to exactly one of A, B:

$$A \triangle B = \{x \in S \mid x \in A \cup B \land x \notin A \cap B\},$$

pronounced "A delta B".

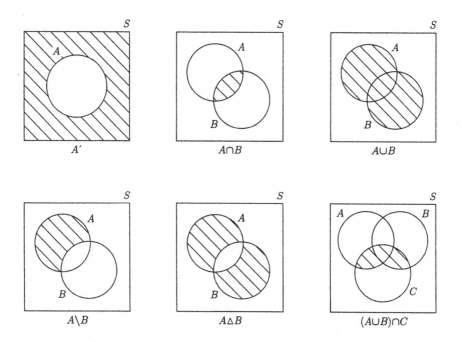

Fig. 3.1. Venn diagrams defining set-theoretic operations.

In some contexts, the terms "meet" and "join" are used in place of intersection and union, respectively, and $A \setminus B$ is sometimes written $A - B$. Sets with no element in common ($A \cap B = \emptyset$) are called **disjoint**. In this case, it is sometimes convenient to write their union as $A \mathbin{\dot\cup} B$, the "disjoint union of A and B".

Operations of this kind can be described in a visual way by means of **Venn diagrams**, in which the prescribed set S is represented by a square in the plane, and subsets of S by circles within this square. Then the elements of S are thought of as points within this square, and the elements of a subset $A \subseteq S$ are thought of as points in the interior of the circle labelled A. Then the five operations defined above are described by hatching the region corresponding to the subset each defines, as in Fig. 3.1. In our analogy between set theory and formal logic, Venn diagrams correspond pretty closely to Lewis Carroll diagrams, and can be used to prove propositions about sets, although an arguably better method will be given in the next section.

To conclude this section, we describe one more construction whereby two given sets may be combined to produce a third. This operation is inherently different from the four binary operations described above, where in each case the two sets to be combined, and also the resulting combinations are subsets of some given set (called S above). The two input sets for this last operation are arbitrary.

Let us start with a familiar example. In coordinate geometry, a point P in the plane is specified by its two coordinates x and y, and we write $P = (x, y)$. The order (x first, y second) is important here: $(1, 2)$ and $(2, 1)$ are different points. So we speak of (x, y) as an **ordered pair**. This is in marked contrast to the braces used to describe sets, where $\{1, 2\} = \{2, 1\}$. Again, we may have an ordered pair (x, y) with $x = y$: $(0, 0)$ is just the origin, for example, whereas to write $\{x, y\}$ when $x = y$ is to abuse notation and is not allowed.

With this in mind, let A and B be arbitrary sets. Then their **Cartesian product** is given by

$$A \times B = \{(a, b) \mid a \in A,\, b \in B\},$$

pronounced "A cross B".

When $A = B$, we get the special case of the Cartesian square $A \times A$, sometimes written $A^{\times 2}$ or just A^2. In the very special case when $A = B = \mathbb{R}$, we get the Cartesian plane $\mathbb{R}^2$. We can similarly define real 3-space $\mathbb{R}^3$ as the set of all ordered triples (x, y, z), $x, y, z \in \mathbb{R}$, which forms the setting for the Newtonian model of the cosmos. The inductive definition

$$\mathbb{R}^1 = \mathbb{R}, \qquad \mathbb{R}^n = \mathbb{R}^{n-1} \times \mathbb{R} \quad \text{for } n > 1$$

gives real n-space, $n \in \mathbb{N}$, whose elements are all n-tuples $(x_1, x_2, \ldots, x_n)$ of real numbers. $\mathbb{R}^n$ is the basic object of study in linear algebra.

EXERCISES

3.15 Exactly one of the four operations $\cup$, $\cap$, $\setminus$, $\triangle$ is non-commutative. Say which one and give a counter-example to the commutative law for it.

3.16 Show that each of the operations $\cup$, $\cap$ on pairs of subsets of a set S has an identity, but inverses do not always exist for these operations when $S \neq \emptyset$.

3.17 Two of the binary operations

$$(A' \cup B) \cap (A \cup B'), \quad (A \setminus B) \cup (B \setminus A), \quad (A' \cup B')', \quad (A \cup B) \setminus (A \cap B)$$

on the subsets of S are the same. By drawing Venn diagrams, say which two.

3.18 Prove the de Morgan laws (cf. Exercise 2.15)

$$(A \cup B)' = A' \cap B', \quad (A \cap B)' = A' \cup B'.$$

3.19 Prove the distributive law (cf. Exercise 2.16)

$$A \cap (B \cup C) = (A \cap B) \cup (A \cap C).$$

3.20 Given disjoint finite sets A and B, prove that

$$|A \cup B| = |A| + |B|.$$

3.21 Given arbitrary finite sets $A, B \subseteq S$, prove that

$$|A \cup B| = |A| + |B| - |A \cap B|.$$

3.22 Let A_1, A_2, A_3 be subsets of a given set S such that $|A_1| = 10$, $|A_2| = 15$, $|A_3| = 20$, $|A_1 \cap A_2| = 8$, $|A_2 \cap A_3| = 9$. By judicious use of the previous exercise, prove that the only values possible for $|A_1 \cup A_2 \cup A_3|$ are 26, 27, 28.

3.23 Given arbitrary finite sets A and B, prove that

$$|A \times B| = |A| \cdot |B|.$$

3.24 Given a finite set A, prove that for all $n \in \mathbb{N}$

$$|A^n| = |A|^n.$$

3.25 Given sets S, T and subsets $C \subseteq S$ and $A, B \subseteq T$, prove the distributive law

$$C \times (A \cup B) = (C \times A) \cup (C \times B).$$

3.3 Laws

The operations of complementation, union, intersection, difference and symmetric difference apply to subsets of a given set S to yield new subsets of S. More complicated operations are obtained by combining them in various ways. When two such combinations result in the *same* new subset for all possible input sets, we have a set-theoretical law. Such **laws**, like those of arithmetic and logic, are statements about sets that are always true. There are nine examples to be found in Exercises 3.15–3.19 in the previous section, such as the distributive law (Exercise 3.19.):

$$A \cap (B \cup C) = (A \cap B) \cup (A \cap C), \qquad (3.4)$$

$\forall A, B, C \subseteq S$.

In every case, the definitions of the set-theoretical operations can be used to convert such a law into a logical law. Thus (3.4) becomes

$$(x \in A) \wedge (x \in B \vee x \in C) \equiv (x \in A \wedge x \in B) \vee (x \in A \wedge x \in C), \qquad (3.5)$$

which can be established using a truth table. Truth tables can therefore be used to prove laws, and also to define operations, in set theory in the same way as in logic. When doing this, we suppress the repetitive "$x \in$" (which occurs seven times in (3.5)), and interpret a 1 or 0 under the symbol for a subset as meaning $x \in$ or $x \notin$ that subset in the row in question. Similarly, a 1 or 0 under a set-theoretical operation means that $x \in$ or $x \notin$ the subset resulting from the application of that operation. The rows correspond to the different basic regions of the Venn diagram, eight in the case of a three-variable law like (3.5), and the corresponding input-tuples of zeros and ones are entered consistently at the outset.

All this will become a lot clearer with the aid of an example. We shall prove the other distributive law, which, though it resembles (3.4), is strangely less intuitive:

$$A \cup (B \cap C) = (A \cup B) \cap (A \cup C). \qquad (3.6)$$

In Table 3.1 we define $\cup$ and $\cap$ in this formulation.

Table 3.1. Definitions by truth table.

A	$\cup$	B
1	1	1
1	1	0
0	1	1
0	0	0

A	$\cap$	B
1	1	1
1	0	0
0	0	1
0	0	0

The third row of the first table thus asserts that: when $x \notin A$, and $x \in B$, $x \in A \cup B$. The truth table for (3.6) is Table 3.2.

Table 3.2. Use of a truth table to prove a law.

A	∪	(B	∩	C)	=	(A	∪	B)	∩	(A	∪	C)
1	1	1	1	1	1	1	1	1	1	1	1	1
1	1	1	0	0	1	1	1	1	1	1	1	0
1	1	0	0	1	1	1	1	0	1	1	1	1
1	1	0	0	0	1	1	1	0	1	1	1	0
0	1	1	1	1	1	0	1	1	1	0	1	1
0	0	1	0	0	1	0	1	1	0	0	0	0
0	0	0	0	1	1	0	0	0	0	0	1	1
0	0	0	0	0	1	0	0	0	0	0	0	0
1	5	2	4	3	6	1	4	2	5	1	4	3

The columns are filled in as follows:

1 A is the column octuple 11110000,

2 B is the column octuple 11001100,

3 C is the column octuple 10101010,

4 the operations within brackets are evaluated first,

5 then the other two, both using Table 3.1, and

6 the relation = is verified by the octuple 11111111,

which appears because the two columns labelled 5 are the same.

As the final example in this section, we verify that the subsets of a given set S form a **group** under the operation of symmetric difference.

Theorem 3.2

Under the binary operation $\triangle$, the subsets of S form (i) an associative system that (ii) has an identity, with respect to which (iii) every element has an inverse.

Proof

Referring to Table 3.3, the first table gives the definition of $\triangle$, and the other two prove that $\emptyset$ is the identity and A is its own inverse, respectively, which verifies (ii) and (iii).

The truth of (i) follows from Table 3.4. □

Table 3.3.

A	$\triangle$	B
1	0	1
1	1	0
0	1	1
0	0	0

A	$\triangle$	$\emptyset$	$=$	A
1	1	0	1	1
0	0	0	1	0

A	$\triangle$	A	$=$	$\emptyset$
1	0	1	1	0
0	0	0	1	0

Table 3.4.

A	$\triangle$	$(B$	$\triangle$	$C)$	$=$	$(A$	$\triangle$	$B)$	$\triangle$	C
1	1	1	0	1	1	1	0	1	1	1
1	0	1	1	0	1	1	0	1	0	0
1	0	0	1	1	1	1	1	0	0	1
1	1	0	0	0	1	1	1	0	1	0
0	0	1	0	1	1	0	1	1	0	1
0	1	1	1	0	1	0	1	1	1	0
0	1	0	1	1	1	0	0	0	1	1
0	0	0	0	0	1	0	0	0	0	0

EXERCISES

3.26 The four unary operations of propositional calculus (see Exercise 2.13) correspond to four unary operations on subsets A of a given set S. What are the four subsets so obtained?

3.27 Use a truth table to prove the law

$$(A \setminus B) \cup (B \setminus A) = (A \cup B) \setminus (A \cap B).$$

3.28 Use a truth table to prove the distributive law (3.4).

3.29 Use Exercise 3.21 in conjunction with the previous exercise to prove that for any subsets $A_1, A_2, A_3 \subseteq S$,

$$\begin{aligned}
|A_1 \cup A_2 \cup A_3| &= |A_1| + |A_2| + |A_3| - |A_1 \cap A_2| - |A_2 \cap A_3| \\
&\quad - |A_1 \cap A_3| + |A_1 \cap A_2 \cap A_3|.
\end{aligned}$$

3.30 Use the analogy between set theory and propositional calculus to draw up truth tables defining the relations $A = B$ and $A \subseteq B$ between subsets $A, B \subseteq S$.

3.31 Use the previous exercise to prove that

$$(A = B) \equiv (A \subseteq B) \wedge (B \subseteq A).$$

3.32 The (almost obvious) laws

$$A \cup A = A, \qquad A \cap A = A$$

are called **idempotent laws** for $\cup$, $\cap$. Of the 16 binary operations on subsets of a set S (cf. Exercise 2.22), how many satisfy the idempotent law?

3.33 Under how many of the 16 binary operations on the subsets of a set S do the latter form a group (cf. Theorem 3.2)?

3.4 The Power Set

It is sometimes necessary to work with sets whose members are themselves sets. Indeed, such a set appeared implicitly in Theorem 3.2 at the end of the previous section, where we proved that the subsets of a given set S form a group under the binary operation $\triangle$. This group is thus a set whose elements are sets. Let us make the following definition.

Definition 3.1

Given any set S, the set of all its subsets is called the **power set** of S, written $\mathcal{P}(S)$.

Following the educational principle that it's easier to understand something once you know how big it is, we elevate Exercise 3.14 to the status of a theorem and, for the sale of variety, give an alternative proof.

Theorem 3.3

If S is a finite set with $|S| = n$, then $|\mathcal{P}(S)| = 2^n$.

Proof

Let the elements of S be $x_1, x_2, \ldots, x_n$. Then any subset $A \subseteq S$ has a kind of "signature" $s(A)$ constructed as follows: $s(A)$ is the binary number of n digits whose kth digit is 1 if $x_k \in A$ and 0 if $x_k \notin A$. So, for example, $s(\emptyset)$ is a string of zeros, $s(S)$ is a string of ones, and in general there are exactly $|A|$ ones in $s(A)$. Since

(a) different subsets have different signatures and

(b) every n-digit binary number is the signature of some subset,

it follows that the total number $|\mathcal{P}(S)|$ of subsets is equal to the total number of n-digit binary numbers, namely 2^n.   □

A subset A of S can thus be thought of in two different ways, as a subset of S or as an element of $\mathcal{P}(S)$:

$$A \subseteq S \Leftrightarrow A \in \mathcal{P}(S).$$

There is scope for confusion here, which increases when we consider, as we sometimes must, *subsets* of $\mathcal{P}(S)$:

$$\sigma \subseteq \mathcal{P}(S) \Leftrightarrow \sigma \in \mathcal{P}(\mathcal{P}(S)).$$

In an attempt to keep things clear, we shall consistently use lower-case Greek letters for subsets of $\mathcal{P}(S)$, and refer to such a subset as a **family** of subsets of S. The rest of this section is given over to the description of two important examples of such families.

A family π of non-empty subsets of a given set S is called a **partition** of S if each element $x \in S$ belongs to exactly one subset $P \in \pi$. This condition can be stated more precisely in terms of symbols:

$$\forall x \in S \; \exists| \; P \in \pi \quad x \in P, \tag{3.7}$$

where the symbol $\exists|$ is pronounced "there is exactly one".

Examples of partitions have already appeared in this chapter. Thus, when $\emptyset \subset A \subset S$, the family $\{A, A'\}$ partitions S into two subsets. Again, when $A, B \subseteq S$, the family $\{A \cap B, A \setminus B, B \setminus A, (A \cup B)'\}$ partitions S into four subsets, provided they are all non-empty ($\emptyset$ is precluded to avoid triviality).

A more concrete example is provided by the set $\mathbb{Z}$ of all integers. Fix a positive integer $n \geq 2$ and consider the set of all its integer multiples,

$$n\mathbb{Z} = \{nk \mid k \in \mathbb{Z}\}.$$

By adding 1 to each element of this set, we get another,

$$1 + n\mathbb{Z} = \{1 + nk \mid k \in \mathbb{Z}\}.$$

Continuing in this way, we get some more,

$$r + n\mathbb{Z} = \{r + nk \mid k \in \mathbb{Z}\},$$

where $r = 2, 3, \ldots, n - 1$. We stop here as $n + n\mathbb{Z} = \mathbb{Z}$ again. A little thought should convince you that, for any given integer m, $m \in r + n\mathbb{Z}$, where r is the (smallest, non-negative) remainder on division of m by $n : m = qn + r$,

$0 \leq r \leq n-1$ (see Theorem 1.14). Since r is uniquely determined by m and n, it follows that every integer belongs to exactly one of the sets $r+n\mathbb{Z}$, $0 \leq r \leq n-1$, which therefore partition $\mathbb{Z}$. This r is called the **residue** of m modulo n, and the sets $r + n\mathbb{Z}$, $0 \leq r \leq n - 1$, are called **residue classes modulo** m. We shall return to this important partition in the next chapter.

To get an example of a partition into infinitely many subsets, take $S = \mathbb{R}$, the real line, and

$$\pi = \{[n, n + 1) \mid n \in \mathbb{Z}\}, \tag{3.8}$$

where the **half-open interval** $[a, b)$, with $a, b \in \mathbb{R}$ and $a < b$, is given by

$$[a, b) = \{x \in \mathbb{R} \mid a \leq x < b\}.$$

Returning to the general theory, it is often convenient to formulate condition (3.7) in terms of members of π alone, without reference to elements x of S. Thus, the fact that every $x \in S$ belongs to some member P of π can be expressed in the form

$$\bigcup_{P \in \pi} P = S. \tag{3.9}$$

The fact that there is only one such P for each x is expressed by the statement

$$(P, Q \in \pi) \wedge (P \neq Q) \Rightarrow P \cap Q = \emptyset \tag{3.10}$$

Conditions (3.9) and (3.10) say respectively that the members of π "cover" S and "do not overlap". The Venn diagram depicting a partition π of S thus looks like a patchwork quilt, as illustrated in Fig. 3.2 with $|\pi| = 7$. We have shown that if π is a partition of S, then (3.9) and (3.10) hold. The converse of this statement (Exercise 3.34) is almost obvious.

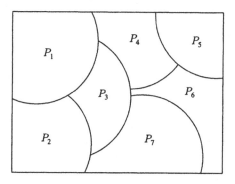

Fig. 3.2. Partitions look like patchwork quilts.

One final important point is as follows. When working with a partition π of a set S, it is often convenient to have at hand a *set T of representatives* of the members of π. Such a subset $T \subseteq S$ forms a **transversal** for π and is defined by the property that in T there is exactly one element of each member of π:

$$\forall P \in \pi \quad |P \cap T| = 1.$$

The single element of $P \cap T$ thus "represents" P in T. The following proposition, although it is a bone of contention among logicians, is accepted almost unanimously by members of the mathematical community.

Definition 3.2

The Axiom of Choice. Every partition of every set has a transversal.

This famous axiom manifests itself in many branches of pure mathematics. It appears in many guises, and at least one of these will occur later in this book.

Our second important example of a family of subsets of a given set S is of an entirely different kind. Though not as ubiquitous as the notion of partition, it is of fundamental importance in that branch of mathematics known as *analysis*. We introduce it by giving a concrete example.

Given real numbers a, b with $a < b$, define the **open interval**

$$(a, b) = \{x \in \mathbb{R} \mid a < x < b\}.$$

Thus, (a, b) is a subset of the real line $\mathbb{R}$, and so is any union of such intervals. Call a union of open intervals an **open subset** of $\mathbb{R}$. Finally, put

$$\tau = \{U \subseteq \mathbb{R} \mid U \text{ is open}\}.$$

We collect some properties of τ into a theorem.

Theorem 3.4

The family τ of open subsets of $\mathbb{R}$ has the following properties:

(i) $\emptyset \in \tau$,

(ii) $\mathbb{R} \in \tau$,

(iii) any union of members of τ is a member of τ,

(iv) $U, V \in \tau \Rightarrow U \cap V \in \tau$.

Proof

(i) $\emptyset$ is the "empty union" (cf. empty sum 0 and empty product 1).

(ii) $\mathbb{R} = \bigcup_{n \in \mathbb{N}} (-n, n)$.

(iii) A union of unions is a union.

(iv) This one is a bit tricky so we'll busk it. The intersection of two unions is a double union of individual intersections (generalized distributive law), and the intersection of two open intervals is again an open interval (possibly empty).   □

Properties (i)–(iv), apart from being a source of exercises on unions and intersections, actually form a system of *axioms*. For any set S, a subset $\tau \in \mathcal{P}(S)$ that satisfies (i)–(iv) (with S in place of $\mathbb{R}$ in (ii)) is called a **topology** on S. A set S equipped with a topology τ is called a **topological space**. The study of topological spaces forms a large and increasingly important area of modern mathematics. We confine ourselves here to a few simple examples in the exercises.

EXERCISES

3.34 Prove that if a family π of non-empty subsets of a set S satisfies (3.9) and (3.10) then π is a partition of S.

3.35 How many partitions are there of the set $S = \{1, 2, 3\}$?

3.36 Write down an upper bound for the number p_n of partitions of a set S with n elements, that is, an expression $e(n)$ depending on n such that $p_n \leq e(n) \; \forall n \in \mathbb{N}$.

3.37 Find a transversal T for the partition π in (3.8).

3.38 A **partition** of a positive integer n is an expression of n as a sum of positive integers. (The order of the terms is ignored $3 + 2$ and $2 + 3$ are the same partition of 5.) If p_n denotes the number of partitions of n (so that $p_2 = 2 : 2$ and $1 + 1$), evaluate p_3, p_4, p_5, p_6. Guess the value of p_7 and then prove your guess is wrong.

3.39 How many topologies are there on the set $\{1, 2\}$?

3.40 Let S be any infinite set and consider the family

$$\kappa = \{A \subseteq S \mid A' \text{ is finite}\} \cup \{\phi\}.$$

Prove that κ satisfies conditions (i)–(iv) of Theorem 3.4. (κ is called the *cofinite topology* on S.)

3.41 A subset C of a topological space S with topology τ is called **closed** if C' is open, that is, $C' \in \tau$. Reformulate conditions (i)–(iv) of Theorem 3.4 in terms of the family $\sigma = \{C \subseteq S \mid C' \in \tau\}$ of closed sets.

3.42 The collection $\mathcal{T}$ of all topologies τ on a set S is a family of subsets of $\mathcal{P}(S)$ (hence, a subset of $\mathcal{P}(\mathcal{P}(S))$ and an element of $\mathcal{P}(\mathcal{P}(\mathcal{P}(S)))$!). Does $\mathcal{T}$ form a topology of $\mathcal{P}(S)$ or not?

4
Relations

Various relations exist between things of the same kind. Thus, if $a, b \in P = \{$all living human beings$\}$, the relation σ of sisterhood may be expressed by writing $a \, \sigma \, b$ if a is a sister of b. Again, the relation $|$ of divisibility exists on the set $\mathbb{Z}$ of all integers: $a \mid b$ if a divides b. In general, a given relation ρ on a set S may or may not hold between two elements $a, b \in S$. Listing or otherwise describing the ordered pairs (a, b) for which it does hold defines the relation. This leads to the following definition, which is a nice example of how to attach a precise mathematical meaning to an everyday word.

Definition 4.1

A (binary) **relation** ρ on a set S is a subset of $S \times S : \rho \subseteq S \times S$. We often write $a \, \rho \, b$ instead of $(a, b) \in \rho$.

This very general definition encompasses two particularly important but very different special types of relation, equivalence relations and orderings, which we will study in turn in this chapter.

4.1 Equivalence Relations

We shall start by defining three properties that a particular relation ρ on a given set S may or may not have. Such a relation is called

$$\begin{array}{llll}
reflexive & \text{(R)} & \text{if} & \forall a \in S \quad a \rho a, \\
symmetric & \text{(S)} & \text{if} & a \rho b \Rightarrow b \rho a, \\
transitive & \text{(T)} & \text{if} & a \rho b \wedge b \rho c \Rightarrow a \rho c,
\end{array}$$

where a, b and c denote elements of S.

To illustrate these general properties, we extend the meagre list of examples given above,

1 sisterhood on P,

2 divisibility on $\mathbb{Z}$,

by adding a few more.

3 Equality on any set S is defined by the **diagonal subset**

$$\Delta = \{(s,s) \mid s \in S\} \subseteq S \times S;$$

$s = t \Leftrightarrow (s,t) \in \Delta$.

4 For $a, b \in \mathbb{R}$, write $a \le b$ if $b - a$ is non-negative.

5 For $a, b \in \mathbb{Z}$, write $a \equiv b$ if $a - b$ is a multiple of 5.

6 For any set S and $A, B \in \mathcal{P}(S)$, write $A \nsubseteq B$ if A is not contained in B.

7 Let S be any set and π any partition of S. Then write $a \equiv b$ for $a, b \in S$ if b belongs to the same member P of π as a.

8 Given $(a, b), (c, d) \in \mathbb{N} \times \mathbb{N}$, write $(a, b) \sim (c, d)$ if $a + d = c + b$.

9 Given $(a, b), (c, d) \in \mathbb{Z} \times \mathbb{N}$, write $(a, b) \sim (c, d)$ if $ad = cb$.

10 Let $\mathbb{R}[x]$ denote the set of all polynomials in an indeterminate x with real coefficients. Then for $f(x), g(x) \in \mathbb{R}[x]$, write $f(x) \sim g(x)$ if $f(x) - g(x) = (x^2 + 1)h(x)$ for some $h(x) \in \mathbb{R}[x]$.

Table 4.1 shows which of these ten relations have which of these three properties. The first seven rows are more or less obvious, and the last three are exercises to which we shall return in Section 3 below. We shall also refer again to row 5 in the next section and to row 7 in a moment, but first, we will look at the main definition.

Table 4.1. Properties of some relations.

	relation	set	R	S	T
1	σ	P			✓
2	$\mid$	$\mathbb{Z}$	✓		✓
3	$=$	S	✓	✓	✓
4	$\leq$	$\mathbb{R}$	✓		✓
5	$\equiv$	$\mathbb{Z}$	✓	✓	✓
6	$\not\subseteq$	$\mathcal{P}(S)$			
7	$\equiv$	S	✓	✓	✓
8	$\sim$	$\mathbb{N} \times \mathbb{N}$	✓	✓	✓
9	$\sim$	$\mathbb{Z} \times \mathbb{N}$	✓	✓	✓
10	$\sim$	$\mathbb{R}[x]$	✓	✓	✓

Definition 4.2

A relation ρ on a set S that is reflexive, symmetric and transitive is called an **equivalence relation** on S.

It is pretty clear that any partition π of a set S determines an equivalence relation on S (in accordance with row 7 of Table 4.1, where it is written $\equiv$). Less obvious but more important is the converse of this statement, and we now begin to work towards the derivation of this fact.

So let $\sim$ denote a fixed equivalence relation on a given set S. Then for any $a \in S$ the set

$$[a] = \{b \in S \mid b \sim a\}$$

is called the **equivalence class** of a. Of course $[a]$ depends on $\sim$ as well as a, and $[a] \subseteq S$. We will prove two facts about equivalence classes. First, every element $a \in S$ belongs to at least one equivalence class: $a = a \in [a]$ because $\sim$ is reflexive. Second, no element $a \in S$ can belong to two different equivalence classes, that is, for $a, b, c \in S$,

$$b \in [a] \wedge b \in [c] \Rightarrow [a] = [c]. \tag{4.1}$$

This is proved using the symmetry and transitivity of $\sim$ as follows. First,

$$b \in [a] \quad \Rightarrow \quad b \sim a, \text{ by definition, and}$$
$$b \in [c] \quad \Rightarrow \quad b \sim c \Rightarrow c \sim b \text{ by (S)},$$

so that $b \sim a$ and $c \sim b$, and $c \sim a$ by (T). Then, using (T) again,

$$d \in [c] \Rightarrow d \sim c \Rightarrow d \sim a \Rightarrow d \in [a],$$

and we have shown that every element d of $[c]$ belongs to $[a]$, that is, $[c] \subseteq [a]$. Finally, by reversing the roles of a and c in the above argument, we obtain $[a] \subseteq [c]$, so that $[a] = [c]$ as claimed.

The two facts we have just established respectively assert that every element of S belongs to at least one equivalence class and to at most one equivalence class, that is, to *exactly one*, and we have proved the following theorem.

Theorem 4.1

If $\sim$ is an equivalence relation on a set S, then the distinct equivalence classes

$$[a] = \{b \in S \mid b \sim a\}, \qquad a \in S,$$

form a partition of S. $\qquad\qquad\qquad\qquad\qquad\qquad\qquad\qquad\qquad\qquad$ □

It turns out that if our equivalence relation $\sim$ comes from a partition π of S, then the resulting equivalence classes are just the members of π. Conversely, starting from an equivalence relation $\sim$, if π is the resulting partition into equivalence classes, then the equivalence relation determined by π is just $\sim$. We thus have a *one-to-one correspondence* between partitions of S and equivalence relations on S. (The notion of one-to-one correspondence, or bijection, will be made precise in the next chapter.) The point is that equivalence relations are easy to set up, whereas partitions are useful, so that Theorem 4.1 is important for practical reasons. Two such applications will be described in the next two sections.

EXERCISES

4.1 Given a set S, which of the properties (R), (S), (T) hold for the relation $a \neq b$ of inequality?

4.2 Which of the properties (R), (S), (T) hold for the relation $\subseteq$ of containment on $\mathcal{P}(S)$?

4.3 Prove that the relation $\sim$ on $\mathbb{N} \times \mathbb{N}$ given in Example 8 is an equivalence relation.

4.4 Do the same for the relation $\sim$ on $\mathbb{Z} \times \mathbb{N}$ in Example 9.

4.5 The same again for the relation $\sim$ on $\mathbb{R}[x]$ in Example 10.

4.6 What is wrong with following "proof" that property (R) of a relation ρ is a consequence of properties (S) and (T)?

First, $a \, \rho \, b \Rightarrow b \, \rho \, a$ by (S) . Then $a \, \rho \, b$ and $b \, \rho \, a \Rightarrow a \sim a$ by (T). Hence ρ is reflexive.

4.7 Any set S admits the trivial partition τ into singletons:

$$\tau = \{\{s\} \mid s \in S\}.$$

Describe the corresponding equivalence relation.

4.8 Describe the partition obtained in accordance with Theorem 4.1 from the equivalence $\equiv$ of Example 5.

4.9 For $x \in \mathbb{R}$, define the **integer part** $[x]$ to be the largest integer $\leq x$. (So $[\pi] = 3$, $[2] = 2$, $[-\frac{1}{2}] = -1$.) Prove that the relation

$$x \sim y \Leftrightarrow [x] = [y]$$

is an equivalence relation on $\mathbb{R}$, describe the equivalence classes and write down a transversal for them. [Note. There is ambiguity of notation here, which unfortunately is traditional. For the purposes of this question, use (x) for the equivalence class of x in $\mathbb{R}$.]

4.10 Write down an upper bound for the number of equivalence classes on a finite set with n elements.

4.11 Prove that the condition

$$a \equiv b \Leftrightarrow a - b \in \mathbb{Z}$$

defines an equivalence relation on the set $\mathbb{Q}$ of rational numbers. Find a transversal T for the equivalence classes.

4.2 Congruences

We focus attention on the relation $\equiv$ in Example 5 of the previous section. Since it depends crucially on the positive integer 5, we incorporate this into our notation and write it more precisely as follows: for $a, b \in \mathbb{Z}$,

$$a \equiv b \pmod 5 \Leftrightarrow 5 \mid (a - b),$$

where the left-hand part is pronounced "a is congruent to b modulo 5" and referred to as a **congruence** (modulo 5).

Let us illustrate Theorem 4.1 in the previous section by working out what the equivalence classes are (as in the solution of Exercise 4.8). For a given $a \in \mathbb{Z}$, divide a by 5 in accordance with Theorem 1.14 to get

$$a = 5q + r, \qquad 0 \leq r \leq 4.$$

Thus, $a \equiv r \pmod{5}$ and so

$$a \in [r], \qquad 0 \le r \le 4,$$

and there are exactly five equivalence classes:

$$[0] = 5\mathbb{Z}, \ [1] = 1 + 5\mathbb{Z}, \ [2] = 2 + 5\mathbb{Z}, \ [3] = 3 + 5\mathbb{Z}, \ [4] = 4 + 5\mathbb{Z},$$

which is just the partition into residue classes modulo 5 (see Section 3.4). The fact that these five classes are pairwise disjoint is expressed mathematically as follows:

$$k, l \in \mathbb{Z} \quad 0 \le k, l \le 4, \qquad k \equiv l \pmod{5} \Rightarrow k = l.$$

It turns out to be possible to add and multiply these classes very much as if they were ordinary individual numbers:

$$[k] + [l] = [k + l], \qquad [k]\,[l] = [kl]. \qquad (4.2)$$

But notice that there is something (rather subtle) here that requires proof. Take the first equation in (4.2). To define $[k] + [l]$, we *choose* elements k, l from each class, add them up, and take the residue class of their sum. But if we made different choices, say $k' \in [k]$ and $l' \in [l]$, how do we know that the result $[k' + l']$ is the same as $[k + l]$? What has to be checked is that the operations in (4.2) are *well-defined*, that is, independent of the choice of representatives from $[k]$ and $[l]$.

That this is indeed the case is expressed mathematically as follows:

$$k' \in [k] \wedge l' \in [l] \Rightarrow [k' + l'] = [k + l] \wedge [k'\, l'] = [k, l],$$

or, more conveniently in terms of congruences,

$$k' \equiv k \pmod{5} \wedge l' \equiv l \pmod{5} \Rightarrow k' + l' \equiv k + l \pmod{5} \wedge k'\, l' \equiv kl \pmod{5},$$

$$(4.3)$$

and we shall prove this now.

Since our hypothesis assert that $k' - k$ and $l' - l$ are both divisible by 5, we can write

$$k' = k + 5a, \quad l' = l + 5b, \qquad a, b \in \mathbb{Z}.$$

Then

$$k' + l' = (k + l) + 5(a + b), k'\, l' = kl + 5(kb + la + 5ab).$$

Since these numbers differ by a multiple of 5 from $k + l$, kl respectively, the conclusion of (4.3) follows.

There is an obvious transversal $T = \{0, 1, 2, 3, 4\}$ for these classes. To avoid the tiresome square brackets, we sometimes replace each class by its representative in T. When this is done, the addition and multiplication tables in

Table 4.2. Addition and multiplication of integers modulo 5.

+	0	1	2	3	4
0	0	1	2	3	4
1	1	2	3	4	0
2	2	3	4	0	1
3	3	4	0	1	2
4	4	0	1	2	3

·	0	1	2	3	4
0	0	0	0	0	0
1	0	1	2	3	4
2	0	2	4	1	3
3	0	3	1	4	2
4	0	4	3	2	1

arithmetic modulo 5 are as shown in Table 4.2. Thus, the entries in the $(2,4)$ places respectively assert that

$$2 + 4 \equiv 1 \pmod 5, \quad 2.4 \equiv 3 \pmod 5.$$

All that we have said so far generalizes nicely from the case $n = 5$ to that of an arbitrary positive integer n (assumed greater than 1 to avoid triviality). In particular, the formulae in (4.2) can be taken as the definition of the sum and product of the residue classes $[k], [l]$ modulo n, $0 \le k, l \le n - 1$. Thus, we can take $T = \{k \in \mathbb{Z} \mid 0 \le k \le n - 1\}$ as a transversal and then the definitions of addition and multiplication modulo n take the form

$$k + l = r, \qquad kl = s,$$

where r, s are the remainders $0 \le r, s \le n - 1$ on dividing $k + l$, kl respectively by n. Independence of choice of representatives is checked in the same way as for $n = 5$ above (see Exercise 4.1) and the resulting set of n residue classes is denoted by $\mathbb{Z}/n\mathbb{Z}$, the "integers modulo n":

$$\mathbb{Z}/n\mathbb{Z} = \{[k] \mid 0 \le k \le n - 1\}.$$

Here $[k]$ denotes the residue class of k modulo n, often abbreviated to k for convenience. We can now state the following theorem, whose proof is left to the exercises (Exercises 4.12–4.16).

Theorem 4.2

(i) The operations of addition and multiplication modulo n are well defined.

(ii) The set $\mathbb{Z}/n\mathbb{Z}$ forms a commutative ring-with-1 under these operations.

(iii) $\mathbb{Z}/n\mathbb{Z}$ is a field if and only if n is prime.   □

What this theorem means in practical terms is that residue classes modulo n can be treated in much the same way as individual numbers in arithmetic. Congruences play the role of equations and are manipulated in a similar way.

The resulting theory of congruences forms an important part of elementary number theory.

To conclude this section, we observe that the type of partition just described can be generalized from $\mathbb{Z}$ to any group G, where the role of $n\mathbb{Z}$ is played by an arbitrary subgroup H of G. The members of the corresponding partition are called **cosets** in this context, and they play a fundamental role in group theory. A word of caution, however, is appropriate in this general case. While the cosets always partition the group, they do not always inherit the group operation. If G is abelian, the natural multiplication of cosets is well defined, but if it isn't the subgroup H needs to be of a special kind, namely, a *normal* subgroup.

EXERCISES

4.12 Show that the sum and product of residue classes $[k], [l]$ modulo n are well defined by the formula (4.2).

4.13 Prove that $\mathbb{Z}/n\mathbb{Z}$ inherits from $\mathbb{Z}$ the five universal laws:

$$(a + b) + c = a + (b + c), \quad a + b = b + a,$$
$$(ab)c = a(bc), \quad ab = ba, \quad a(b + c) = ab + ac,$$

in the eight axioms for a ring-with-1.

4.14 Prove that $\mathbb{Z}/n\mathbb{Z}$ satisfies the three existential laws:

$$\exists \text{ additive identity}, \ \exists \text{ multiplicative identity},$$
$$\exists \text{ additive inverse for every element},$$

in the eight axioms for a ring-with-1.

4.15 Show that if n is not a prime, then $\mathbb{Z}/n\mathbb{Z}$ is not a field.

4.16 Use Theorem 1.15 to show that $\mathbb{Z}/n\mathbb{Z}$ is a field when n is a prime.

4.17 Let $[a]$ be a residue class modulo n, $0 \le a \le n - 1$. Show that $[a]$ has a multiplicative inverse in $\mathbb{Z}/n\mathbb{Z}$ if and only if a and n are coprime, $(a, n) = 1$.

4.18 What are the residue classes modulo 2 better known as?

4.19 Let A be the residue class of 1 modulo 3 and B that of -1 modulo 4. Prove that $A \cap B$ is a residue class modulo 12.

4.3 Number Systems

In the previous section, an equivalence relation was used to construct a new number system from an old one: the relation on $\mathbb{Z}$ of congruence modulo n led to the construction of $\mathbb{Z}/n\mathbb{Z}$. We continue this theme now, but with the difference that the new systems we construct will be *bigger* than the old ones. In each case, the new system will contain the old one and will enjoy the advantage of containing the solutions to more equations than did the old one. We shall construct three number systems using Examples 8, 9, 10 from Section 1 in turn.

Recall Example 8, where the equivalence relation $\sim$ is defined on $\mathbb{N} \times \mathbb{N}$ by

$$(a, b) \sim (c, d) \Leftrightarrow a + d = c + b. \tag{4.4}$$

The old number system here is $\mathbb{N}$, and the new one consists of the equivalence classes $[(a, b)]$ of this relation, which we will abbreviate to $[a, b]$ for the sake of convenience.

First notice that we can subtract any $k \in \mathbb{N}$ from both a and b to get an equivalent pair, provided $k < \min(a, b)$:

$$(a, b) \sim (a - k, b - k).$$

It follows that every class $[a, b]$ contains a special pair, that is, one in which at least one of the components is equal to 1. Moreover, there is only one such pair in each class:

$$(1, b) = (1, d) \Rightarrow b = d, \ (a, 1) = (b, 1) \Rightarrow a = b, \ (a, 1) = (1, b) \Rightarrow a = b = 1.$$

The special pairs thus form a transversal,

$$T = \{(1, n + 1) \mid n \in \mathbb{N}\} \cup \{(1, 1)\} \cup \{(n + 1, 1) \mid n \in \mathbb{N}\}. \tag{4.5}$$

Next, attempt to define the addition of classes in a natural way:

$$[a, b] + [c, d] = [a + c, b + d]. \tag{4.6}$$

As in the previous section, we need to check for independence of representatives (Exercise 4.20) in order that this operation be well defined. The effect of this operation on the members of T works out as follows.

$$\forall (a, b) \in \mathbb{N} \times \mathbb{N}, \ [a, b] + [1, 1] = [a + 1, b + 1] = [a, b] \tag{4.7}$$

so that $[1, 1]$ is an *identity* for this $+$. Next,

$$\begin{aligned}
[m + 1, 1] + [n + 1, 1] &= [m + n + 2, 2] = [(m + n) + 1, 1], \\
[1, m + 1] + [1, n + 1] &= [2, m + n + 2] = [1, (m + n) + 1].
\end{aligned} \tag{4.8}$$

so that the first and last sets on the right-hand side of (4.5) both look like copies on $\mathbb{N}$. Finally,

$$[n+1,1] + [1,n+1] = [n+2,n+2] = [1,1], \tag{4.9}$$

so that $[1,n+1]$ is the *additive inverse* of $[n+1,1]$.

The set of equivalence classes forms a number system you may already recognize. It has the following properties:

(a) its elements can be added (4.6),

(b) it contains a copy of $\mathbb{N}$ (4.8),

(c) it also contains an additive identity $[1,1]$ (4.7),

(d) the remaining elements are the additive inverses of those in (b) (4.9).

So the classes form the set $\mathbb{Z}$ of all integers, where we usually write

$$[n+1,1] = n, \qquad [1,1] = 0, \qquad [1,n+1] = -n.$$

This may seem like a lot of trouble to take just to construct familiar objects like negative integers. But bear in mind that the Ancient Romans had no zero, and negative numbers were beyond the scope of human intellect in mediæval times. Their existence was in those days the cause of controversy, and belief in them required an act of faith. Thanks to set theory, that act of faith can now be replaced by an act of reason, as described above. There may be quicker ways of constructing $\mathbb{Z}$, but the one we have just described is the cleanest.

$\mathbb{Z}$ is a richer number system than $\mathbb{N}$ in the following sense. The equation

$$x + b = a \tag{4.10}$$

with $a, b \in \mathbb{N}$ can be solved for $x \in \mathbb{N}$ only when $b < a$. Allowing $x \in \mathbb{Z}$, it can be solved for all $a, b \in \mathbb{N}$:

$$x = [a,b] = [a+2,b+2] = [a+1,1] + [1,b+1] = a - b.$$

Moreover, since $a - b$ has a meaning for all $a, b \in \mathbb{Z}$ (by Exercise 4.21, $-(-b) = b$), (4.10) has a solution in $\mathbb{Z}$ for all $a, b \in \mathbb{Z}$.

The set $\mathbb{Z}$ has more structure, namely, *multiplication*:

$$[a,b]\,[c,d] = [ac+bd, ad+bc], \tag{4.11}$$

which, like addition, can be shown to be well defined. Under these two operations, $\mathbb{Z}$ forms a commutative ring-with-1. It is a routine calculation to check the eight axioms; some of these appear in the exercises. Finally, $\mathbb{Z}$ has no divisors of zero (Exercise 4.24), and so is an *integral domain*.

We now turn to Example 9 in Section 1, which we treat in much the same way, but more briefly. The relation $\sim$ is defined on $\mathbb{Z} \times \mathbb{N}$ by

$$(a, b) \sim (c, d) \Leftrightarrow ad = cb. \tag{4.12}$$

First notice that if k is a common divisor of a and b ($k \in \mathbb{Z}$, $k \geq 2$), say $a = ka'$ and $b = kb'$, then

$$(a, b) \sim (a', b').$$

It follows that every class $[a, b]$ contains a pair with coprime components. Moreover, there is only one such pair in each class: suppose that

$$(a, b) \sim (c, d), \text{ that is, } ad = cb,$$

with a, b coprime and c, d coprime. Then it follows from the fundamental theorem of arithmetic (Theorem 1.17) that $a = c$ and $b = d$.

Next, attempt to define multiplication of classes in the natural way:

$$[a, b][c, d] = [ac, bd]. \tag{4.13}$$

As in the case of (4.6) above, this is independent of choice of representatives (Exercise 4.25). The following facts are then a matter for routine checking:

(a) $\{[a, 1] \mid a \in \mathbb{Z}\}$ is a copy of the integers $\mathbb{Z}$,

(b) $[1, 1]$ is an identity for the multiplication (4.13),

(c) if $a > 0$, $[a, b]$ has a multiplicative inverse $[b, a]$, and if $a < 0$, the inverse of $[a, b]$ is $[-b, -a]$,

(d) the addition of classes is well defined by the rule

$$[a, b] + [c, d] = [ad + cb, bd], \tag{4.14}$$

(e) $[0, 1] = \{(0, b) \mid b \in \mathbb{N}\}$ is the additive identity,

(f) $[-a, b]$ is the additive inverse of $[a, b]$,

(g) the classes $[a, b]$ have all the familiar properties of the rational numbers a/b, $a \in \mathbb{Z}$, $b \in \mathbb{N}$, a and b coprime.

The above process thus extends the integers $\mathbb{Z}$ to the rational numbers $\mathbb{Q}$. $\mathbb{Q}$ is a field, and any equation $ax = b$ with $a, b \in \mathbb{Q}$ and $a \neq 0$ has a solution $x \in \mathbb{Q}$.

Finally, we turn to Example 10 in Section 1, where the analysis has points in common with that in Section 2 above. The relation $\sim$ on $\mathbb{R}[x]$ is defined by

$$f(x) \sim g(x) \Leftrightarrow (x^2 + 1) \mid (f(x) - g(x)). \tag{4.15}$$

By the analogue for $R[x]$ of Euclid's theorem (Theorem 1.14), dividing any polynomial $f(x)$ by $x^2 + 1$ leaves a remainder $r(x)$ of degree < 2:

$$f(x) = (x^2 + 1)\, q(x) + r(x).$$

Then $f(x) \sim r(x)$ and every equivalence class contains a linear polynomial $r(x) = a + bx$, $a, b \in \mathbb{R}$. Moreover, since the only multiple of $x^2 + 1$ of degree less than 2 is the zero polynomial, no class can contain more than one such polynomial. It follows that the linear polynomials form a transversal for the classes,

$$T = \{a + bx \mid a, b \in \mathbb{R}\}.$$

Classes can be added and multiplied in a natural way:

$$[f(x)] + [g(x)] = [f(x) + g(x)], \qquad [f(x)]\,[g(x)] = [f(x)\, g(x)], \qquad (4.16)$$

by means of the usual addition and multiplication of polynomials. The following facts can then be established:

(a) the classes form a *field* under these operations,

(b) the classes $[a]$ of constant polynomials $a \in \mathbb{R}$ form a copy of the real numbers $\mathbb{R}$,

(c) the class $[x]$ satisfies the equation

$$[x]^2 + [1] = [x^2 + 1] = [0],$$

where $[1], [0]$ are the multiplicative, additive identities respectively,

(d) the classes $[a + bx]$ have all the familiar properties of the complex numbers $a + i\,b$, $a, b \in \mathbb{R}$.

This process thus extends the reals $\mathbb{R}$ to the field $\mathbb{C}$ of complex numbers. We saw earlier that any non-constant linear polynomial $ax - b \in \mathbb{Q}[x]$, $a \neq 0$, has a zero $x = b/a \in \mathbb{Q}$. $\mathbb{C}$ has the vastly stronger property of **algebraic closure**: any non-constant polynomial $f(x) \in \mathbb{C}[x]$ has a zero in $\mathbb{C}$. This is the fundamental theorem of algebra; its proof is beyond our scope.

Theorem 4.3

The field $\mathbb{C}$ of complex numbers is algebraically closed.   □

Of all number systems, the richest are $\mathbb{R}$ and $\mathbb{C}$. While $\mathbb{C}$ has the advantage of algebraic closure, $\mathbb{R}$ is superior in having a natural *ordering*: given $a, b \in \mathbb{R}$, either $a < b$ or $a = b$ or $a > b$ (more on this in the next section). This ordering is lost in the passage to $\mathbb{C}$, and two further examples of such a loss are as follows. The complex numbers $\mathbb{C}$ can in turn be extended to a bigger system $\mathbb{H}$, the quaternions of Hamilton, but in the process the commutative law of multiplication is lost. Again, $\mathbb{H}$ can be extended to the Cayley numbers $\mathbb{A}$, but at the expense of the associative law of multiplication. As far as I know, the story ends here, and we have a nice chain of number systems,

$$\mathbb{N} \subseteq \mathbb{Z} \subseteq \mathbb{Q} \subseteq \mathbb{R} \subseteq \mathbb{C} \subseteq \mathbb{H} \subseteq \mathbb{A}. \tag{4.17}$$

So far, we have said nothing about:

(a) the extension of $\mathbb{Q}$ to $\mathbb{R}$,

(b) the existence of $\mathbb{N}$.

Of course, you can't create something out of nothing, but in the case of (b) we can get away with remarkably little: the existence of $\mathbb{N}$, its three elements of structure $(+, \cdot, \leq)$ and all their properties can be deduced from the three axioms of Peano, and this will be discussed in the next chapter.

On the other hand, the construction of the reals $\mathbb{R}$ from the rationals $\mathbb{C}$ is a non-trivial business, and was not convincingly managed until the last century. And then, surprisingly, it was achieved independently by two very different methods. The common feature of these methods is that they both employ partitions, and we describe the key to each of them now.

First, the method of **Dedekind sections** constructs real numbers as partitions $\{A, B\}$ of the rational numbers $\mathbb{Q}$ with the property that

$$\forall a \in A, b \in B \qquad a < b. \tag{4.18}$$

Since this condition endows the pair A, B with an order, we shall denote this partition by (A, B). For example, for any $c \in \mathbb{Q}$ the sets

$$A = \{q \in \mathbb{Q} \mid q \leq c\}, \qquad B = \{q \in \mathbb{Q} \mid q > c\}$$

form a partition of $\mathbb{Q}$. So do the sets

$$C = \{q \in \mathbb{Q} \mid q < c\}, \qquad D = \{q \in \mathbb{Q} \mid q \geq c\},$$

and both of these partitions correspond to the rational number q. But if A has no largest element and B has no smallest element, the partition (A, B) corresponds to an irrational real number. It is clear that, for two partitions (A, B) and (C, D), $A \subseteq C \Leftrightarrow B \supseteq D$, and in this case we write $(A, B) \leq (C, D)$. Addition and multiplication admit natural definitions in a similar way, and all the familiar properties of $\mathbb{R}$ can be developed in a logical manner. Especially important in this context is the **Dedekind property**: every bounded set of real numbers has a least upper bound.

The second method constructs real numbers as equivalence classes of **Cauchy sequences**. These are sequences $a_1, a_2, \ldots, a_n, \ldots$ of rational numbers whose terms "eventually become arbitrarily close". More precisely, this means that

$$\forall k \in \mathbb{N} \; \exists N \in \mathbb{N} \quad m, n > N \Rightarrow -\frac{1}{k} < a_m - a_n < \frac{1}{k}, \qquad (4.19)$$

or in English: no matter how small a positive rational number $1/k$ you choose, there is some point (the Nth term) of the sequence beyond which any two terms differ by less than $1/k$. Such a sequence is said to be **convergent** to a **limit** $l \in \mathbb{Q}$ if the terms get arbitrarily close to l in the same sense. A **null sequence** is a convergent sequence whose limit $l = 0$. Finally, two Cauchy sequences are **equivalent** if they differ by a null sequence. The resulting equivalence classes then turn out to have all the structure and properties of the field $\mathbb{R}$ of real numbers. Especially important in this context is the property of **completeness**: every Cauchy sequence of real numbers is convergent.

As you might guess, the study of the real numbers belongs properly in analysis.

EXERCISES

4.20 Prove that the addition of classes given by (4.6) is well defined.

4.21 Interpreting $-(-b)$ as $(-1)(-b)$ for $b \in \mathbb{N}$, use (4.11) to prove that $-(-b) = b$.

4.22 Prove that the multiplication (4.11) is distributive over the addition (4.6).

4.23 Prove that $[2, 1]$ is a multiplicative identity for $\mathbb{Z}$.

4.24 By distinguishing four cases and using the fact that $mn + 1 \neq 1 \; \forall m, n \in \mathbb{N}$ prove that $\mathbb{Z}$ has no divisors of zero.

4.25 Prove that the multiplication of classes given by (4.13) is well defined.

4.26 Prove that $[-a, b]$ is the additive inverse $[a, b]$ according to (4.14).

4.27 Use the definitions (4.16) to add and multiply the complex numbers $[a + bx], [c + dx]$ and translate the resulting formulae into the usual notation.

4.28 If $c + id$ is the multiplicative inverse of the non-zero complex number $a + ib$, express c, d in terms of a, b.

4.29 Write down natural definitions for the sum and product of two Dedekind sections (A, B) and (C, D).

4.30 For each $k \in \mathbb{N}$, let d_k be an integer with $0 \le d_k \le 9$. Prove that the sequence

$$a_n = \sum_{i=1}^{n} d_i 10^{-i}, \qquad n \in \mathbb{N},$$

is a Cauchy sequence. Express in words the relationship between the d_k and the real number r corresponding to the equivalence class $[a_n]$ of the sequence (a_n).

4.4 Orderings

Whereas equivalence relations may be the most useful type of relation, orderings are arguably the most fundamental. Orderings come in different strengths, and we define these now.

Definition 4.3

A **partial ordering** on a set S is a relation ρ on S with the following three properties:

(R) ρ is reflexive, $\forall a \in S \ a \, \rho \, a$,

(O) for $a, b \in S$, $a \, \rho \, b \wedge b \, \rho \, a \Rightarrow a = b$,

(T) ρ is transitive, for $a, b, c \in S$, $a \, \rho \, b \wedge a \, \rho \, b \Rightarrow a \, \rho \, c$.

A partial ordering which satisfies the extra condition

(L) $\forall a, b \in S \ a \, \rho \, b \vee b \, \rho \, a$

is called a **total ordering** (or a linear ordering).

A partial ordering which satisfies the extra condition

(W) every non-empty subset of S has a least element with respect to ρ, that is,

$$\emptyset \ne A \subseteq S \Rightarrow \exists l \in A \ \forall a \in A \ l \, \rho \, a,$$

is called a **well-ordering**.

Of the ten examples of relations given in Section 1, just three are partial orderings, namely, relations 2, 3, 4 of divisibility on $\mathbb{Z}$, equality on S, $\le$ on $\mathbb{R}$ respectively. None of these (with the trivial exception of the second when $|S| \le 1$)

satisfies either (L) or (W). More examples will be given shortly, but we first
make some comments on these three definitions in a series of remarks.

1 The superficial resemblance between the definitions of equivalence relation
  and partial ordering is misleading. These two types of relation have almost
  nothing in common (cf. Exercise 4.31).

2 A relation ρ on a set S satisfying (R) and (T) is sometimes called a **quasi-ordering**.

3 Conspicuous in its failure to comply with our definition of partial ordering in the relation $<$ on $\mathbb{R}$. This is merely a matter of convention (see
  Exercise 4.32): the relation

  $$a \leq b \Leftrightarrow a < b \vee a = b$$

  is a partial ordering on $\mathbb{R}$.

4 The relation $\leq$ is even a total ordering on $\mathbb{R}$ (but not a well ordering).
  In terms of the relation $<$, this is equivalent to the **law of trichotomy**,
  which states that $\forall a, b \in \mathbb{R}$, exactly one of the following three conditions
  holds:

  $$a < b, \qquad a = b, \qquad b < a.$$

5 Condition (W) on a partial ordering ρ is genuinely stronger than condition
  (L):

  $$(W) \Rightarrow (L)$$

  but not conversely in general (Exercise 4.33).

6 The well-ordering principle mentioned in Chapter 1 merely asserts that the
  relation $\leq$ is a well-ordering on $\mathbb{N}$.

7 The Axiom of Choice holds for well-ordered sets (see Exercise 4.35). Conversely, it follows from the Axiom of Choice that every set has a well-
  ordering, but the proof of this fact is beyond our scope.

8 Certainly, every *finite* set A can be well ordered. Simply list the elements
  of A in some order: $a_1, a_2, \ldots, a_n$, where $n = |A|$, and define $a_i \leq a_j \Leftrightarrow i \leq j$.

We conclude this section with a brief discussion of some especially important
examples of orderings of numbers and words.

Prompted by Remark 3 above, define a partial ordering $\leq$ on $\mathbb{N}$ in two
stages as follows:

$$\begin{aligned}
a < b \quad &\Leftrightarrow \quad \exists x \in \mathbb{N} \; a + x = b, \\
a \leq b \quad &\Leftrightarrow \quad a < b \vee a = b.
\end{aligned} \qquad (4.20)$$

This can be extended in a natural way to partial orderings on $\mathbb{Z}$ and $\mathbb{Q}$ (Exercise 4.36) and even $\mathbb{R}$ (Exercise 4.38) in terms of the constructions described in the previous section. Neither $\mathbb{Z}$ nor the set of positive rationals is well ordered by this relation (Exercise 4.37).

Given a finite set A, $|A| = n \in \mathbb{N}$ say, a **word** in A is a string of symbols $w = x_1 x_2 \ldots x_l$ with each $x_i \in A$, $1 \leq i \leq l$, that is, an element of A^l with the brackets and commas omitted. l is called the **length** $l(w)$ of w, and we allow the empty word e of length 0. In this context, A is often referred to as an **alphabet**. Thus, the number of words of length l in A is n^l for all $l \geq 0$. As in Remark 8, listing the elements of A as $a_1, a_2, \ldots, a_n$ defines a well-ordering on A.

The set W of all words in A can be well ordered in two different ways, of which the first is as follows. Given $u, v \in W$, say

$$u = x_1 \ldots x_m, \qquad v = y_1 \ldots y_n, \qquad (4.21)$$

write $u < v$ if either

(a) at the first place from the left (say the ith) where u and v differ, $x_i < y_i$, or

(b) there is no such place and $m < n$.

Then $u \leq v \Leftrightarrow u < v \vee u = v$ is called the **lexicographic ordering** on W.

Another well-ordering on W is obtained by tampering with the previous definition as follows given $u, v \in W$ as in (4.21) write $u < v$ if either

(a) $m < n$, or

(b) $m = n$ and, at the first place (say the ith) where u and v differ, $x_i < y_i$,

and again put $u \leq v \Leftrightarrow u < v \vee u = v$. This is called the **shortlex ordering** of W. This ordering plays a fundamental role in such areas as computer science and combinatorial group theory.

EXERCISES

4.31 What can you say about a relation ρ on a set S that is both a partial ordering and an equivalence relation?

4.32 Let σ be a relation on a set S that is transitive (T) and irreflexive (I), $\forall a \in S \sim (a \, \sigma \, a)$. Prove that the relation $a \, \rho \, b \Leftrightarrow a \, \sigma \, b \vee a = b$ is a partial ordering on S.

4.33 Let ρ be a partial ordering on S. Prove that: ρ satisfies (W) $\Rightarrow \rho$ satisfies (L). Show by finding a counter-example that the converse is false in general.

4.34 Find counter-examples to prove that none of the relations of divisibility on $\mathbb{Z}$, equality on S, containment on $\mathcal{P}(S)$ is a total ordering (for $|S| > 1$).

4.35 Let π be a partition of a well-ordered set S. Describe a transversal T for π.

4.36 Extend the definition (4.20) to define partial orderings on $\mathbb{Z}$ and $\mathbb{Q}$.

4.37 Prove that neither $\mathbb{Z}$ nor the positive rationals is well ordered by the natural orderings of the previous exercise.

4.38 How might you extend the natural ordering from $\mathbb{Q}$ to $\mathbb{R}$ using Cauchy sequences?

4.39 Let W be the set of words in the alphabet $A = \{0, 1\}$. Then the subset

$$S = \{x_1 \ldots x_m \in W \mid m \geq 1, x_1 = 1\}$$

is just the set $\mathbb{N}$ of positive integers in binary notation. Compare the natural, lexicographic and shortlex oderings on S.

4.40 With A and W as in the previous exercise, and

$$T = \{x_1 \ldots x_m \in W \mid m \geq 1, x_m = 1\},$$

the set $.T = \{.w \mid w \in T\}$ is just the set of dyadic rationals in $(0, 1)$ in binary form. Compare the natural, lexicographic and shortlex orderings on $.T$.

5
Maps

"He considered that the most important and necessary part
of the study of geography was the drawing of maps ... "
A. Chekhov
The Teacher of Literature

The third fundamental notion of set theory, after *set* itself and *relation*, is that
of a *map*. Whereas a relation involves only one set and we speak of a *relation
on a set A*, a map involves two and we speak of a *map between sets A and B*.
The sets A and B here play different roles, and to emphasize this we often
speak of *maps from A to B*.

The word map has many synonyms including mapping, function, operation, transformation, operator and morphism, depending on the context. The
French word for map is "application" and the German word is "Abbildung",
and "mapping" is an accurate translation of the latter.

5.1 Terminology and Notation

Although the notion of map can, like that of relation, be defined in terms of
more basic concepts (like subset and Cartesian product as below), both the
traditional use and our intuitive perception of the word involve the idea of an
action on the elements of one set to produce elements of another.

Given set A and B, a **map** θ **from** A **to** B is a rule that assigns to each
element $a \in A$ an element of B, called the **image** of a under θ and written
$\theta(a)$. It is customary to express this by writing

$$\theta : A \to B,$$

and to visualize θ as an agent that acts on each element $a \in A$ to produce an element $\theta(a) \in B$.

The set A is called the **domain** of θ, and B the **codomain**. Given an equation $\theta(a) = b$ with $a \in A$ and $b \in B$, we sometimes refer to b as the **value** of θ on a, and to a as an **argument** of θ. The set of values of θ on all $a \in A$ is a subset of B called the **image** of θ,

$$\operatorname{Im} \theta = \{b \in B \mid \exists a \in A \; \theta(a) = b\}. \tag{5.1}$$

Two maps are said to be **equal** if they

(a) have the same domain,

(b) have the same codomain, and

(c) take the same value on every element of the domain.

Thus, if $\theta : A \to B$ and $\phi : C \to D$ are two maps, then

$$\theta = \phi \Leftrightarrow (A = C) \wedge (B = D) \wedge (\forall a \in A \; \theta(a) = \phi(a)).$$

Given sets A and B, a particular map $\theta : A \to B$ is determined by its values $\theta(a) \; \forall a \in A$. It is sometimes possible to specify θ by writing

$$\begin{array}{rcl} \theta : \quad A & \to & B \\ a & \mapsto & b, \end{array} \tag{5.2}$$

where $b \in B$ is some expression depending on a that defines the value of θ on a. Alternatively, θ may be specified by listing the pairs (a, b) with $\theta(a) = b$ $\forall a \in A$, that is, by specifying the subset

$$S = \{(a, b) \mid b = \theta(a)\} \subseteq A \times B. \tag{5.3}$$

Thus, a map from A to B can be thought of (and even *defined*) as a subset $S \subseteq A \times B$ with the property that

$$\forall a \in A \; \exists | \; b \in B \; (a, b) \in S, \tag{5.4}$$

whereupon the unique b is written $\theta(a)$ and called the value of θ on a.

While a subset of $A \times B$ satisfying (5.4) can be used as the definition of the term "map from A to B", it is customary to take the definition given above, and to refer to the set S in (5.3) as the **graph** of θ.

Given two sets A and B, it is possible to consider all maps from A to B together as the elements of a third set, written $\operatorname{Map}(A, B)$. The old notation B^A, now seldom used, for the set $\operatorname{Map}(A, B)$ is justified by the following fact.

Theorem 5.1

Let A and B be finite sets, so that $|A| = a$ and $|B| = b$ are non-negative integers. Then the cardinality of $\mathrm{Map}(A, B)$ is given by

$$|\mathrm{Map}(A, B)| = b^a.$$

Proof

A map $\theta : A \to B$ determines and is determined by its value on each element $x \in A$. For each such x there are b possibilities for $\theta(x)$ and there are a such x's. So θ is specified by making a choice of one out of b things (in b ways) a times independently. So the total number of maps is just b^a. $\qquad\square$

Two maps θ and ϕ such that *the domain of ϕ coincides with the codomain of θ* can be combined to produce a third map $\phi\theta$ ($\phi \circ \theta$ in some books) called the **composite** of θ and ϕ and defined as follows. If $\theta : A \to B$ and $\phi : B \to C$, then

$$\phi\theta : \quad A \;\to\; C$$
$$a \;\mapsto\; \phi(\theta(a)).$$

This composition of maps is itself a map

$$\mathrm{Map}(A, B) \times \mathrm{Map}(B, C) \to \mathrm{Map}(A, C).$$

Note that, because we are writing maps on the *left* of their arguments, $\phi\theta$ is the result of applying θ first, then ϕ. In many areas of algebra, maps are written on the *right* (so $\theta(x)$ is written $x\theta$) to avoid this difficulty.

A very useful, but almost obvious, fact is as follows.

Theorem 5.2

Composition of maps is associative.

Proof

Let $\theta : A \to B$, $\phi : B \to C$, $\psi : C \to D$.

Given $a \in A$, put

$$\theta(a) = b, \quad \phi(b) = c, \quad \psi(c) = d.$$

Then

$$(\phi\,\theta)(a) = c \Rightarrow (\psi(\phi\,\theta))(a) = \psi(c) = d,$$

and

$$(\psi\,\phi)(b) = d \Rightarrow ((\psi\,\phi)\theta)(a) = (\psi\phi)(b) = d.$$

Since $\psi(\phi\,\theta)$ and $(\psi\,\phi)\theta$ have the same domain (A), the same codomain (D) and take the same value (d) on a for all $a \in A$, these maps are equal. $\square$

For any set X, define the **identity map on** X by setting

$$1_X : \quad X \quad \rightarrow \quad X$$
$$x \quad \mapsto \quad x,$$

the map that fixes every element. Then it is clear that, for any map $\theta : A \rightarrow B$,

$$1_B\theta = \theta = \theta 1_A.$$

Since we have identity maps in $\mathrm{Map}(A, B)$, it is natural to ask about the existence of inverses. Although this question cannot be answered fully until Section 3 below, we can at least say something.

Given a map $\theta : A \rightarrow B$, we can associate with any element $b \in B$ the set of those elements of A that take the value b under θ. This set is called the **pre-image** of b under θ and written $\theta^{-1}(b)$, pronounced "θ-inverse of b". Thus, $\forall b \in B$

$$\theta^{-1}(b) = \{a \in A \mid \theta(a) = b\}. \tag{5.5}$$

The notation $\theta^{-1}(b)$ suggests that θ^{-1} is a map with domain B. But unless we are very lucky (see Section 3 below), the codomain of θ^{-1} is *not* A. Equation (5.5) shows that the values of θ^{-1} are actually *subsets* of A, whence the codomain of θ^{-1} is the power set $\mathcal{P}(A)$,

$$\theta^{-1} : B \rightarrow \mathcal{P}(A). \tag{5.6}$$

More generally, we can make the following definition:

$$\forall Y \subseteq B \ \theta^{-1}(Y) = \{a \in A \mid \theta(a) \in Y\}, \tag{5.7}$$

of the pre-image of any subset $Y \subseteq B$. This delivers a map

$$\theta^{-1} : \mathcal{P}(B) \rightarrow \mathcal{P}(A), \tag{5.8}$$

of which the map (5.6) is an important special case. The map in (5.8) will prove useful later on (Section 6.2 below).

EXERCISES

5.1 Give two reasons why "taking square-roots" is *not* a map from $\mathbb{R}$ to $\mathbb{R}$.

5.2 Why did we not define the image of $\theta : A \rightarrow B$ more simply as $\{\theta(a) \mid a \in A\}$?

5.3 Let $A = \{a, b\}$ and $B = \{1, 2, 3\}$. By describing their graphs, list the elements of Map(A, B).

5.4 With A, B as in the previous exercise, list the elements of Map(B, A).

5.5 For given sets A and B, our definition of "graph" shows that this is itself a map, call it γ. What are the domain, codomain and image of γ?

5.6 Given non-empty sets A and B, describe Map$(A, \emptyset)$ and Map$(\emptyset, B)$. What about Map$(\emptyset, \emptyset)$?

5.7 Let $\theta : A \to B$, $\phi : B \to C$ be maps with graphs G, H respectively. Describe the graph K of $\phi\theta$ in terms of G and H.

5.8 Given a map $\theta : A \to B$, prove that the family $\theta^{-1}(b)$, $b \in \operatorname{Im}\theta$, forms a partition of A.

5.9 Let $\theta : A \to B$ be a map. Bearing in mind your solution to Exercise 5.2 above, make a natural definition of a map $\theta : \mathcal{P}(A) \to \mathcal{P}(B)$. To avoid awkwardness, a harmless abuse of notation is being committed here, as in formula (5.8) in the text. What is this abuse and what makes it harmless?

5.10 Let $\theta : A \to B$ be a map and regard θ and θ^{-1} as maps between power sets as in the previous exercise and formula (5.8) in the text, respectively. Of the four proposed laws

$$\forall X, Y \subseteq A \quad \theta(X \cup Y) = \theta(X) \cup \theta(Y),$$
$$\theta(X \cap Y) = \theta(X) \cap \theta(Y),$$
$$\forall X, Y \subseteq B \quad \theta^{-1}(X \cup Y) = \theta^{-1}(X) \cup \theta^{-1}(Y),$$
$$\theta^{-1}(X \cap Y) = \theta^{-1}(X) \cap \theta^{-1}(Y),$$

exactly three are true. Find a counter-example in your solution to Exercise 5.3 that invalidates the false one.

5.2 Examples

It is high time we had some concrete examples of maps. Since maps are very
numerous (Theorem 5.1 above), we shall need to be selective, and so will choose
examples to illustrate a number of important special types of map. We treat
six of these in turn, and leave others as exercises.

5.2.1 Inclusions

These maps form a convenient way of expressing the containment of one set in
another. If B is any set and A is any subset of B, then the map

$$\iota(A,B): \begin{array}{ccc} A & \to & B \\ a & \mapsto & a \end{array} \tag{5.9}$$

is called the **inclusion** of A in B. It is usual to write simply ι (or inc) for this
map when the domain and codomain are clear from the context.

For any A and B with $A \subseteq B$, there is just one inclusion $\iota : A \to B$, even
in the case when $A = \phi$. At the other extreme, when $A = B$, the inclusion is
just the identity map 1_B defined in the previous section.

Suppose we have, in addition to the inclusion $\iota : A \to B$, an arbitrary map
$\theta : B \to C$. Since the codomain of ι and the domain of θ are equal (to B),
ι and θ can be composed in accordance with the definition in the previous
section. The resulting map $\theta\iota : A \to C$ is called the **restriction** of θ to A,
often written $\theta|_A$. Changing the emphasis, let $\iota : A \to B$ be inclusion and
$\phi : A \to C$ a map. Then any map $\theta : B \to C$ satisfying $\theta\iota = \phi$ is called an
extension of ϕ to B.

Now let A be any set and $A \times A$ its Cartesian square. While $A \times A$ cannot
have A as a subset, it has a subset that looks very like A, namely the **diagonal
subset**

$$\Delta = \{(a,b) \in A \times A \mid a = b\} = \{(a,a) \mid a \in A\}.$$

We can then define the **diagonal map**

$$\delta_A: \begin{array}{ccc} A & \to & A \times A \\ a & \mapsto & (a,a), \end{array} \tag{5.10}$$

which looks very like an inclusion. It shares with any inclusion ι the important
property that *distinct elements of the domain have distinct images under the
map*:

$$\text{for } a, a' \in A, \quad a \neq a' \Rightarrow \iota(a) \neq \iota(a'). \tag{5.11}$$

5.2.2 Natural Maps

Like inclusions, these maps are of a very general kind. Let A be any set and $\sim$ any equivalence relation on A. Then (by Theorem 4.1) the equivalence classes are members of a partition of A, which is usually written $A/\sim$, pronounced "A modulo tilde". $A/\sim$ is sometimes called the **quotient set** of A with respect to $\sim$. Since every $a \in A$ belongs to exactly one equivalence class $[a] \in A/\sim$, there is a **natural map**

$$\begin{aligned} \nu(A, \sim): \quad A &\to A/\sim \\ a &\mapsto [a]. \end{aligned} \qquad (5.12)$$

The dependence on A and $\sim$ is often suppressed, and we merely write ν (or nat) for this map.

For example, take A to be $\mathbb{Z}$ and $\sim$ to be congruence modulo 6, so that $A/\sim$ is $\mathbb{Z}/6\mathbb{Z}$. Then the natural map

$$\nu : \mathbb{Z} \to \mathbb{Z}/6\mathbb{Z}$$

assigns to each $n \in \mathbb{Z}$ its residue class modulo 6.

We conclude this subsection by giving an example of a map that looks very like a natural map. Given non-empty sets X and Y, there are obvious maps

$$\begin{aligned} \pi_1: \quad X \times Y &\to X, & \pi_2: \quad X \times Y &\to Y \\ (x, y) &\mapsto x, & (x, y) &\mapsto y, \end{aligned}$$

called **projections** onto the first and second components respectively. These maps share with any natural map $\nu : A \to B$ the important property that *their codomain and image coincide*:

$$\forall b \in B \; \exists a \in A \quad \nu(a) = b. \qquad (5.13)$$

5.2.3 Binary Operations

We have seen examples of binary operations already, such as the addition of positive integers

$$\begin{aligned} \mathbb{N} \times \mathbb{N} &\to \mathbb{N} \\ (a, b) &\mapsto a + b \end{aligned}$$

and the union of subsets

$$\begin{aligned} \mathcal{P}(S) \times \mathcal{P}(S) &\to \mathcal{P}(S) \\ (A, B) &\mapsto A \cup B. \end{aligned}$$

In general, a **binary operation** on a set A is just a map

$$\beta : A \times A \to A.$$

The image of such a map is usually written in the form $a * b$, $a, b \in A$, rather than $\beta(a, b)$, where $*$ is the operation concerned, such as $+$ (addition) and $\cdot$ (multiplication) when A is a number system, and $\cup$ and $\cap$ when A is a power set.

5.2.4 Polynomial Maps

The polynomials of elementary algebra are complicated objects. They are generally written in the form

$$f(x) = \sum_{k=0}^{n} c_k x^k, \tag{5.14}$$

where x is a "variable" or "indeterminate", the **coefficients** c_k are "constants", and n is a non-negative integer called the **degree** of $f(x)$ provided $c_n \neq 0$. In the case when the coefficients c_k are allowed to range over the real numbers $\mathbb{R}$, the set of polynomials of the form (5.14), including 0, is usually written $\mathbb{R}[x]$.

A polynomial $f(x) \in \mathbb{R}[x]$ and a real number $a \in \mathbb{R}$ together determine another real number, usually written $f(a)$, obtained by *substituting* a for x in the right-hand side of (5.14):

$$f(a) = \sum_{k=0}^{n} c_k a^k.$$

From this there arise two maps:

(a) for a fixed $a \in \mathbb{R}$, the **substitution map**

$$
\begin{array}{rccc}
r_a : & \mathbb{R}[x] & \to & \mathbb{R} \\
      & f(x) & \mapsto & f(a),
\end{array}
$$

which is important in number theory, and

(b) for a fixed $f(x) \in \mathbb{R}[x]$, the **polynomial map**

$$
\begin{array}{rccc}
f : & \mathbb{R} & \to & \mathbb{R} \\
    & a & \mapsto & f(a),
\end{array}
\tag{5.15}
$$

which is important in analysis.

5.2.5 Pointwise Sum and Product

Let A and B be sets, where B has a binary operation

$$\beta: \quad \begin{aligned} B \times B &\to B \\ (b, b') &\mapsto b * b' \end{aligned}$$

defined on it. Then given two maps θ, $\phi \in \mathrm{Map}(A, B)$, we get another map $\theta * \phi \in \mathrm{Map}(A, B)$ by setting

$$\theta * \phi: \quad \begin{aligned} A &\to B \\ a &\mapsto \theta(a) * \phi(a). \end{aligned} \qquad (5.16)$$

We thus have a binary operation on $\mathrm{Map}(A, B)$ *induced pointwise* by the binary operation $*$ on B,

$$\begin{aligned} \mathrm{Map}(A, B) \times \mathrm{Map}(A, B) &\to \mathrm{Map}(A, B) \\ (\theta, \phi) &\mapsto \theta * \phi. \end{aligned}$$

Of particular importance here is the case $A = B = \mathbb{R}$, and it is usual to refer to maps $\theta, \phi \in \mathrm{Map}(\mathbb{R}, \mathbb{R})$ as *functions*. Then the addition and multiplication on $\mathbb{R}$ induce **pointwise addition and multiplication** on $\mathrm{Map}(\mathbb{R}, \mathbb{R})$:

$$\theta + \phi: \quad \begin{aligned} \mathbb{R} &\to \mathbb{R} \\ a &\mapsto \theta(a) + \phi(a), \end{aligned} \qquad \theta . \phi: \quad \begin{aligned} \mathbb{R} &\to \mathbb{R} \\ a &\mapsto \theta(a)\phi(a). \end{aligned}$$

Note that the notation $\theta . \phi$ is used for the pointwise product to distinguish it from the composite $\theta\phi$ defined in the previous section.

These concepts provide an alternative definition of polynomial maps, as follows. First, define a **constant map** on $\mathbb{R}$ to be one of the form

$$\begin{aligned} \mathbb{R} &\to \mathbb{R} \\ a &\mapsto c, \end{aligned}$$

where c is some *fixed* real number. (When $c = c_0$, this is just the polynomial map f of (5.15) arising from $f(x)$ in (5.14) when $n = 0$.) Next consider the "power maps"

$$\begin{aligned} \mathbb{R} &\to \mathbb{R} \\ a &\mapsto a^n, \end{aligned}$$

where $n \in \mathbb{N}$, obtained from the identity map $1_{\mathbb{R}}$ by iterated pointwise multiplication with itself (defined by induction on n). Then take pointwise products of these with constant maps and, finally, use iterated pointwise summation of the resulting maps to get the general polynomial map (5.15). In this sense, the polynomial maps are just the maps *generated* by $1_{\mathbb{R}}$ and the constant maps under pointwise addition and multiplication.

5.2.6 Elementary Functions

These are the basic objects of study in calculus, and comprise:

(a) the polynomial maps,

(b) the trigonometric functions,

(c) the exponential function, and

(d) all maps generated under composition and pointwise addition, multiplication and division (where possible) by these and their inverses.

It remains to say what the inverse of a map is. For example, the inverses of the maps (functions) x^2, e^x, $\sin x$ are $\sqrt{x}$, $\log x$, $\sin^{-1} x$, respectively. So "inverse" here means "inverse under composition of maps", as described at the end of the last section. But calculus is concerned with maps: from $\mathbb{R}$ to $\mathbb{R}$, not from $\mathbb{R}$ to $\mathcal{P}(\mathbb{R})$, and none of $\sqrt{x}$, $\log x$, $\sin^{-1} x$ qualify as maps in this sense. The fundamental reason why this is so will be explained in the next section.

EXERCISES

5.11 Let A and B be sets with $\emptyset \neq A \subseteq B$ and $\iota : A \to B$ the corresponding inclusion. Prove that there is a map $\sigma : B \to A$ such that $\sigma\iota = 1_A$.

5.12 Let A be any set, $\sim$ any equivalence relation on A, $Q = A/\sim$ the corresponding quotient set, and $\nu : A \to Q$ the natural map. Assuming the Axiom of Choice, prove that there is a map $\tau : Q \to A$ such that $\nu\tau = 1_Q$.

5.13 Let $A, \sim, Q, \nu$ be as in the previous exercise, and $A', \sim', Q', \nu'$ be defined similarly, and let $\theta : A \to A'$ be a map such that $a \sim b \Rightarrow \theta(a) \sim' \theta(b)$ for $a, b \in A$. Prove that there is a map $\theta' : Q \to Q'$ such that $\theta'\nu = \nu'\theta$.

5.14 Again let $A, \sim, Q, \nu$ be as in Exercise 5.12, and let $*$ be a binary operation on A. How might $*$ be used to define a binary operation on Q? Find a condition on $\sim$ and $*$ under which this is possible.

5.15 In addition to ordinary addition and multiplication, at least six binary operations on $\mathbb{N}$ appear in Chapter 1. See how many you can remember (or find).

5.16 Let f be the map (5.15) arising from (5.14). What can you say about $|f^{-1}(0)|$?

5.17 Given two maps $\theta : A \to B$, $\phi : C \to D$, define their **product** by setting

$$\theta \times \phi : \quad A \times C \quad \to \quad B \times D$$
$$(a, c) \quad \mapsto \quad (\theta(a), \phi(b)).$$

In the case when $A = C$, $B = D$ and there is a binary operation $*$ on the latter set, write $\theta * \phi$ as the composite of $\theta \times \phi$ with two other maps.

5.18 Let A be a set with a binary operation $*$, so that the set $\mathrm{Map}(A, A)$ has two binary operations, $*$ and composition defined on it. Write down four candidates for a distributive law and say which of them is true in general.

5.19 What would be a good general name for a member of $\mathrm{Map}(\mathbb{N}, \mathbb{R})$?

5.20 Define the **characteristic function** χ_A of a subset A of a set S by

$$\chi_A : \quad S \quad \to \quad \{0, 1\}$$
$$s \quad \mapsto \quad \begin{cases} 1 & \text{if } s \in A, \\ 0 & \text{if } s \notin A. \end{cases}$$

Express $\chi_{A \cap B}$ and $\chi_{A \triangle B}$ in terms of χ_A and χ_B.

5.3 Injections, Surjections and Bijections

Two kinds of map are of special importance. They are the maps that share the property (5.11) of inclusions and the property (5.13) of natural maps, respectively. We shall discuss each of these properties in turn, and then look at maps which enjoy them both.

Definition 5.1

Injections. A map $\theta : A \to B$ is said to be **injective** (or one-to-one, or an injection), if it takes distinct values on distinct elements of its domain: for $a, a' \in A$,

$$a \neq a' \Rightarrow \theta(a) \neq \theta(a'). \tag{5.17}$$

When checking for injectivity, it is often more convenient to verify the contrapositive: for $a, a' \in A$,

$$\theta(a) = \theta(a') \Rightarrow a = a'. \tag{5.18}$$

It turns out that injections can be characterized in two useful ways: in terms of the subsets $\theta^{-1}(b) \subseteq A$, and in terms of maps: $B \to A$. Since the single element of $\mathrm{Map}(\emptyset, B)$ is vacuously injective, we may assume that $A \neq \emptyset$.

Theorem 5.3

For $\theta \in \mathrm{Map}(A, B)$, $A \neq \emptyset$, the following conditions are equivalent

(a) θ is an injection,

(b) $\forall b \in B \; |\theta^{-1}(b)| \leq 1$,

(c) $\exists \phi \in \mathrm{Map}(B, A) \phi\theta = 1_A$.

Proof

There are six assertions here, but in view of the transitivity of the relation $\Rightarrow$ it is sufficient to prove three of them.

(a) $\Rightarrow$ (b). We shall prove the contrapositive. So suppose $\exists b \in B \; |\theta^{-1}(b)| \geq 2$, say $a, a' \in \theta^{-1}(b)$ with $a \neq a'$. Then $\theta(a) = \theta(a') = b$, and so θ is not an injection.

(b) $\Rightarrow$ (c). By definition, $\mathrm{Im}\,\theta = \{b \in B \mid \theta^{-1}(b) \neq \emptyset\}$. Then $\forall b \in \mathrm{Im}\,\theta$, $\theta^{-1}(b)$ is a singleton, $\{b'\}$ say, $b' \in A$. Since $A \neq \emptyset$, $\exists x \in A$. Then define

$$\phi: \quad B \quad \to \quad A$$
$$b \quad \mapsto \quad \begin{cases} b' & \text{if } b \in \mathrm{Im}\,\theta, \\ x & \text{if } b \notin \mathrm{Im}\,\theta. \end{cases}$$

Then $\forall a \in A$, the single element $\theta(a)'$ of $\theta^{-1}(\theta(a))$ is obviously equal to a. Thus,

$$\forall a \in A, \quad \phi(\theta(a)) = \theta(a)' = a = 1_A(a).$$

(c) $\Rightarrow$ (a). Let $\phi \in \mathrm{Map}(B, A)$ be such that $\phi\theta = 1_A$. Then $\forall a, a' \in A$,

$$\theta(a) = \theta(a') \Rightarrow \phi(\theta(a)) = \phi(\theta(a')) \Rightarrow a = a',$$

and θ is injective by (5.18). $\square$

Definition 5.2

Surjections. A map $\theta : A \to B$ is said to be **surjective** (or onto, or a surjection) if every element of B is the image under θ of some element of A:

$$\forall b \in B \; \exists a \in A \quad \theta(a) = b. \tag{5.19}$$

This might look very different from the condition (5.17) for an injection, but there is a parallel to be drawn between injections and surjections. This is illustrated by the following characterizations of surjective maps, which closely parallel those in Theorem 5.3.

Theorem 5.4

For $\theta \in \text{Map}(A, B)$, the following conditions are equivalent:

(a) θ is a surjection,

(b) $\forall b \in B \; |\theta^{-1}(b)| \geq 1$,

(c) $\exists \psi \in \text{Map}(B, A) \; \theta\psi = 1_B$.

Proof

(a) $\Rightarrow$ (b). Let θ be a surjection. Then it follows from (5.19) that $\forall b \in B$ $\theta^{-1}(b) \neq \emptyset$, that is, $|\theta^{-1}(b)| \geq 1$.

(b) $\Rightarrow$ (c). Assuming (b), there is an element of A, call it b', in $\theta^{-1}(b)$ $\forall b \in B$. Then $\theta(b') = b$, and the map $\psi : B \to A$, $b \mapsto b'$, has the property that

$$\theta(\psi(b)) = \theta(b') = b = 1_B(b),$$

whence $\theta\psi = 1_B$. (Did you spot our appeal to the Axiom of Choice here?)

(c) $\Rightarrow$ (a). Assume (c) and let $b \in B$. Since

$$b = 1_B(b) = \theta(\psi(b)) \in \text{Im} \, \theta,$$

θ is surjective.   □

We end this subsection with a look at the important special cases when A or B (or both) is a finite set.

Let A and B be sets and $\theta \in \text{Map}(A, B)$. When A is finite, say $A = \{a_1, \ldots, a_n\}$, the image of θ consists of the distinct elements of the form $\theta(a_i)$, $1 \leq i \leq n$. But if θ is injective, all these are different and $|\text{Im} \, \theta| = n$. But if θ is not injective, at least two of them are equal and $|\text{Im} \, \theta| < n$. Thus, when A is finite,

$$\theta \text{ is injective} \Leftrightarrow |\text{Im} \, \theta| = |A|. \tag{5.20}$$

Now let B be finite. If θ is surjective, then $\text{Im} \, \theta = B$ and so $|\text{Im} \, \theta| = |B|$. But if θ is not surjective, $\text{Im} \, \theta \subset B$ and $|\text{Im} \, \theta| < |B|$. Thus, when B is finite,

$$\theta \text{ is surjective} \Leftrightarrow |\text{Im} \, \theta| = |B|. \tag{5.21}$$

When A and B are both finite, it follows from (5.20) and (5.21) that any two of the assertions

$$\theta \text{ is injective, } \theta \text{ is surjective, } |A| = |B|$$

imply the third. There is a special case of this that is particularly useful; the proof is obvious.

Theorem 5.5 (pigeonhole principle)

If A and B are finite sets with $|A| = |B|$ and $\theta \in \mathrm{Map}(A, B)$, then

$$\theta \text{ is injective} \Leftrightarrow \theta \text{ is surjective.}$$

Definition 5.3

Bijections. A map $\theta : A \to B$ is called a **bijection** if it is both injective and surjective.

Such maps are particularly important, as we hope to demonstrate now. Given a bijection $\theta : A \to B$, let $\phi, \psi \in \mathrm{Map}(B, A)$ be as guaranteed by Theorems 5.3(c), 5.4(c). Then, using associativity of map composition

$$\psi = 1_A \psi = (\phi\theta)\psi = \phi(\theta\psi) = \phi 1_B = \phi,$$

and we have

$$\phi\theta = 1_A, \quad \theta\phi = 1_B. \tag{5.22}$$

In this situation, it makes sense to call ϕ the **inverse** of θ and to write $\phi = \theta^{-1}$. Note that it follows from the symmetry of equations (5.22) that the inverse ϕ of θ is also a bijection.

The relation of this θ^{-1} to the $\theta^{-1} : B \to \mathcal{P}(A)$ at the end of Section 1 needs some explaining. There is ambiguity here, but not much. For a given bijection $\theta : A \to B$, we deduce from Theorems 5.3(b) and 5.4(b) that $\forall b \in B$ $|\theta^{-1}(b)| = 1$, that is, $\theta^{-1}(b)$ is a singleton. The new $\theta^{-1}(b)$ is thus equal to the single element in the old $\theta^{-1}(b)$. Note that we have reserved use of the *word* "inverse" for the new θ^{-1} only.

The argument leading to (5.22) proves half of the following theorem. The other half follows from the assertions (c) $\Rightarrow$ (a) in Theorems 5.3 and 5.4.

Theorem 5.6

A map $\theta : A \to B$ has an inverse if and only if it is a bijection. $\square$

Bijections from a set to *itself* have a special name: they are called **permutations**. Given a set A, we write

$$\text{Sym}(A) = \{\theta \in \text{Map}(A, A) \mid \theta \text{ is bijective}\}$$

for the set of all permutations of A. The origin of the notation "Sym" will be explained in a moment.

Theorem 5.7

For any set A, $\text{Sym}(A)$ forms a group under composition of maps.

Proof

We check the axioms for a group in turn.

First, composition really is a binary operation; that is, the composite $\theta\phi$ of bijections $\theta, \phi \in \text{Sym}(A)$ is indeed a bijection:

$$(\theta\phi)(\phi^{-1}\theta^{-1}) = \theta(\phi\phi^{-1})\theta^{-1} = \theta\theta^{-1} = 1_A,$$

by the associativity of map composition. Thus, $\theta\phi$ has an inverse, $\phi^{-1}\theta^{-1} \in \text{Sym}(A, A)$, and so is a bijection by Theorem 5.6.

Next, the associative law (which has already been used) is guaranteed by Theorem 5.2.

Finally, the identity is obviously 1_A, and inverses exist by Theorem 5.6 again. □

In virtue of this theorem, we can refer to $\text{Sym}(A)$ as the **symmetric group** on A; this is traditional terminology among group theorists. When A is finite, say $|A| = n \in \mathbb{N}$, $\text{Sym}(A)$ is called the **symmetric group of degree** n, and often written S_n, especially when $A = \{k \in \mathbb{Z} \mid 1 \leq k \leq n\}$.

EXERCISES

5.21 Let $A = \{a, b\}$, $B = \{1, 2, 3\}$ as in Exercise 5.3. How many elements of $\text{Map}(A, B)$ are injections? How many are surjections?

5.22 With A, B as in the previous exercise, how many elements of $\text{Map}(B, A)$ are surjections? How many are injections?

5.23 Let $\theta : \mathbb{R} \to \mathbb{R}$ be the map given by $\theta(x) = ax + b$, where $a, b \in \mathbb{R}$ are fixed. Under what conditions on a, b is θ

(i) an injection,   (ii) a surjection,   (iii) a bijection?

What about polynomial maps of degree ≥ 2?

5.24 Consider the **Euler totient function** $\phi : \mathbb{N} \to \mathbb{Z}$, where $\phi(n) = |\{k \in \mathbb{N} \mid 1 \leq k < n \wedge (k,n) = 1\}|$. Is ϕ injective? Is ϕ surjective?

5.25 Concoct a map $\sigma : \mathbb{N} \to \mathbb{N}$ that is injective but not surjective.

5.26 Concoct a map $\tau : \mathbb{N} \to \mathbb{N}$ that is surjective but not injective.

5.27 Let A and B be finite sets with $|A| = m$ and $|B| = n$, say. How many injections are there from A to B? How many bijections? Why am I not asking you to count the surjections?

5.28 Let $\theta : A \to B$ be any map and Q the partition of A consisting of distinct, non-empty pre-images $\theta^{-1}(b) = \{a \in A \mid \theta(a) = b\}$. Prove that there is a bijection $\theta' : Q \to \operatorname{Im}\theta$.

5.29 Use the result of the previous exercise to express any map $\theta : A \to B$ as the composite of a surjection, a bijection and an injection, in that order.

5.30 Given a set S, use characteristic functions (see Exercise 5.20) to define a bijection from $\mathcal{P}(S)$ to $\operatorname{Map}(S, \{0, 1\})$. What can you now deduce from Theorem 5.1?

5.31 Given a set S, show that the complementation map $\kappa : A \mapsto A'$ on $\mathcal{P}(S)$ is a bijection.

5.32 Given a bijection $\theta : A \to B$, construct a bijection from $\mathcal{P}(A)$ to $\mathcal{P}(B)$.

5.33 Let A and B be disjoint sets (contained in some set S if you like). Find a bijection from $\mathcal{P}(A \cup B)$ to $\mathcal{P}(A) \times \mathcal{P}(B)$.

5.34 Given sets A, B, C with A and B disjoint, find a bijection between $\operatorname{Map}(A \cup B, C)$ and $\operatorname{Map}(A, C) \times \operatorname{Map}(B, C)$.

5.35 Given arbitrary sets A, B, C, find a bijection β between $\operatorname{Map}(A \times B, C)$ and $\operatorname{Map}(A, \operatorname{Map}(B, C))$.

5.4 Peano's Axioms

We briefly postpone further study of bijections in order to describe the wonderfully economical axioms of Peano for the fundamental number system $\mathbb{N}$. Given only three simple elements of structure satisfying only three simple axioms, it is possible to deduce all the familiar properties of $\mathbb{N}$, that is, all the laws satisfied by the two binary operations ($+$ and $\cdot$) and the ordering ($\leq$). The problem of getting the Principle of Mathematical Induction is solved by default: it is included among the axioms.

Postulate 1

There exists a *set N*, an *element* $1 \in N$ and a *map* $\sigma : N \to N$, $\sigma(a) = a^*$, such that the following three conditions hold:

P1. σ is an injection,

P2. $1 \notin \operatorname{Im} \sigma$,

P3. if a subset X of N contains 1, and $a^* \in X$ whenever $a \in X$, then $X = N$.

    A few comments are appropriate. First note that, since σ is injective but not surjective, it follows from the pigeon-hole principle (Exercise 5.36) that N is not a finite set: once $+$, $\cdot$, and $\leq$ have been defined on it, N will later become the familiar set $\mathbb{N}$ of positive integers. Next, the distinguished element 1 will later turn out to be the identity with respect to $\cdot$ and also the smallest element of N with respect to $\leq$. Finally, axiom P3 is recognizable as the PMI; it will be used to make definitions and prove theorems in what follows. As an example of how it is applied, we begin with the following useful little lemma. (A **lemma** is a proposition that is of little interest in itself but useful in proving subsequent theorems.)

Lemma 5.1

1 is the only element of N that is not in $\operatorname{Im} \sigma$.

Proof

The subset $X = \{1\} \cup \operatorname{Im} \sigma \subseteq N$ contains both 1 and the image $a^* = \sigma(a)$ of every $a \in N$, whence of every $a \in X$. It follows from P3 that $X = N$, as required. $\qquad\qquad\square$

Armed with this tool, we now embark on the development of the structure of N.

Definition 5.4

Addition. We define a binary operation $+$ on N by the rules

$$A1. \quad \forall a \in N \quad a + 1 = a^*,$$
$$A2. \quad \forall a, b \in N, \quad a + b^* = (a + b)^*.$$

Theorem 5.8

The operation $+$ is well defined throughout N.

Proof

There are two things to be proved here:

(a) $\forall a, b \in N$ at least one meaning can be attached to $a + b$,

(b) $\forall a, b \in N$ at most one meaning can be attached to $a + b$,

and we treat these in turn.

(a) establishes the "throughout" and is proved as follows. Let

$$X = \{b \in N \mid a + b \text{ has a meaning } \forall a \in N\}.$$

Then $1 \in X$ by A1. Also, $b \in X \Rightarrow b^* \in X$ by A2. It follows from P3 that $X = N$, as required.

(b) asserts that $+$ is unique, that is, it is the only binary operation on N satisfying A1 and A2. This is proved by contradiction as follows. Let $\circ$ be another binary operation on N that satisfies A1 and A2, and put

$$X = \{b \in N \mid a \circ b = a + b \,\forall a \in N\}.$$

Then by A1, $a \circ 1 = a^* = a + 1$, so $1 \in X$. Next, it follows from A2 that

$$\begin{aligned} b \in X \quad &\Rightarrow \quad a + b = a \circ b \,\forall a \in N \quad \Rightarrow \quad (a + b)^* = (a \circ b)^* \,\forall a \in N \\ &\Rightarrow \quad a + b^* = a \circ b^* \,\forall a \in N \quad \Rightarrow \quad b^* \in X. \end{aligned}$$

It now follows from P3 that $X = N$, whence $a + b = a \circ b \,\forall a, b \in N$. This contradiction proves the uniqueness of $+$. $\square$

All assertions are proved in the same fashion: define a suitable set X, prove that $1 \in X$, prove that $b \in X \Rightarrow b^* \in X$, and deduce from P3 that $X = N$. The assertion then follows, either by contradiction or otherwise. The proof of the associative law (Exercise 5.37) is typical, while the commutative law requires two steps (Exercises 5.38 and 5.39) and is a kind of double induction; the cancellation law (Exercise 5.40) is an extension of axiom P1.

Theorem 5.9

The binary operation $+$ on N obeys the following laws:

(a) $(a+b)+c = a+(b+c)$,

(b) $a+b = b+a$,

(c) $a+c = b+c \Rightarrow a = b$.

Proof

(a) Take $X = \{c \in N \mid (a+b)+c = a+(b+c) \; \forall a,b \in N\}$.

(b) First take $X = \{a \in N \mid a+1 = 1+a\}$, then $Y = \{b \in N \mid a+b = b+a \; \forall a \in N\}$.

(c) Take $X = \{c \in N \mid$ the map $a \mapsto a+c$ on N is an injection$\}$. $\qquad \square$

Definition 5.5

Multiplication. We define a binary operation $\cdot$ on N by the rules

$$\begin{array}{lll} \text{M1.} & \forall a \in N & a \cdot 1 = a, \\ \text{M2.} & \forall a,b \in N & a \cdot b^* = a \cdot b + a. \end{array}$$

As usual, we tend to omit the $\cdot$ and write the product $a \cdot b$ simply as ab. This is another recursive definition, and an analogue (Exercise 5.41) of Theorem 5.8 is required. The proofs of the distributive law for $\cdot$ over $+$, and of the associative, commutative and cancellation laws for $\cdot$ are consigned to the exercises.

Theorem 5.10

The following laws hold in N:

(a) $a(b+c) = ab + ac$,

(b) $a(bc) = (ab)c$,

(c) $ab = ba$,

(d) $ac = bc \Rightarrow a = b$.   □

Definition 5.6

Ordering. We define a relation $<$ on N by setting

$$a < b \Leftrightarrow \exists x \in N \quad a + x = b.$$

It is then natural to write, for $a, b \in N$,

$$a \leq b \Leftrightarrow a < b \lor a = b.$$

The latter relation turns out to be a partial ordering on N and (just for a change) we give a proof of this fact.

Theorem 5.11

The relation $\leq$ is a partial ordering on N.

Proof

According to the definition in Section 4.4, we must prove that

(a) $\forall a \in N \ a \leq a$,

(b) $a \leq b \land b \leq a \Rightarrow a = b$,

(c) $a \leq b \land b \leq c \Rightarrow a \leq c$.

First, (a) is obvious: $\forall a \in N \ a = a \Rightarrow a < a \lor a = a \Rightarrow a \leq a$. For (c) observe that if either $a = b$ or $b = c$ or both, the conclusion coincides with one or both of the two premises, and the implication is trivial. There remains the case when $a < b$ and $b < c$. Then $\exists x, y \in N \ a + x = b \land b + y = c$, and we have

$$c = b + y = (a + x) + y = a + (x + y)$$

by the associative law, whence $a < c$, which implies $a \leq c$, as required.

Turning to (b), assume that $a \leq b \ \land \ b \leq a \ \land \ a \neq b$ and look for a contradiction. Since $a \neq b$, we have

$$
\begin{aligned}
a < b \land b < a \ &\Rightarrow \ \exists x, y \in N \quad a + x = b \land b + y = a \\
&\Rightarrow \ a = b + y = (a + x) + y = a + (x + y) \\
&\Rightarrow \ a + 1 = a + (x + y) + 1 \\
&\Rightarrow \ 1 = (x + y) + 1 = (x + y)^*, \text{ using Theorem 5.9(b) and (c),} \\
&\Rightarrow \ 1 \in \operatorname{Im} \sigma,
\end{aligned}
$$

which contradicts axiom P2.   □

Just as $+$ and $\cdot$ are related to each other by a law (the distributive law), they are each related to $\leq$. These relations take the form of implications resembling cancellation laws and their proof is left to the exercises.

Theorem 5.12

For $a, b \in N$, the following are equivalent:

(a) $a < b$,   (b) $\forall c \in N \; a + c < b + c$,   (c) $\exists c \in N \; a + c < b + c$.

Theorem 5.13

For $a, b \in N$, the following are equivalent:

(a) $a < b$,   (b) $\forall c \in N \; ac < bc$,   (c) $\exists c \in N \; ac < bc$.

Finally it turns out that $\leq$ is in fact a well-ordering on N, and we end this chapter with a proof. This will require three preliminary facts, which we collect into a theorem. Further properties of $\leq$ are to be found in the exercises.

Theorem 5.14

(a) $\forall a \in N \; 1 \leq a$.

(b) $\forall a, b \in N \; a < b$ or $a = b$ or $b < a$.

(c) $\forall b \in N \; \nexists a \in N \; b < a < b^*$.

Proof

(a) It follows from the lemma that $\forall a \in N$ either $a = 1$ or $a = \sigma(b) = b^* = b + 1$ for some $b \in N$, and then $1 < a$.

(b) Put $A = \{ b \in N \mid \forall a \in N \; a < b$ or $a = b$ or $b < a \}$. Then $1 \in A$ by part (a). Let $b \in A$ and $a \in N$, so that $a < b$ or $a = b$ or $b < a$. In the first case, $a < b < b^*$ and in the second $a = b < b^*$. In the third case, $a = b + x$ and we distinguish two subcases. Either $x = 1$, when $a = b^*$, or $x \neq 1$. In the second subcase, it follows from part (a) that $x = y + 1$ for some $y \in N$. Then, using results already proved,

$$a = b + x = b + y + 1 = b + 1 + y = b^* + y > b^*.$$

It follows in every case that $b^* \in A$, whence $A = N$, as required.

(c) Suppose, for a contradiction that $b < a < b^*$. Then $b^* = a + x$ and $a = b + y$, whence

$$b + 1 = b^* = a + x = b + y + x \Rightarrow 1 = y + x,$$

by Theorem 5.9(c). Then $x < 1$, which contradicts part (a).   □

Theorem 5.15

The relation $\leq$ is a well-ordering on N.

Proof

Let $A \subseteq N$ with $A \neq \emptyset$ and define

$$B = \{b \in N \mid b \leq a \; \forall a \in A\}.$$

We claim that $B \cap A$ contains just one element, and this is the required least element of A. Since $B \cap A$ cannot contain more than one element (Theorem 5.11(b)), we only need to prove that $A \cap B \neq \emptyset$

Since $A \neq \emptyset$, $\exists a \in A$. Then $a < a^*$, so that $a^* \notin B$. Thus $B \neq N$. By contraposing axiom P3 we must have either

$$1 \notin B \text{ or } \exists b \in B \; b^* \notin B.$$

But $1 \in B$ by Theorem 5.14(a). Thus we have a $b \in B$ with $b^* \notin B$. Then $\exists a \in A$ with $\sim(b^* \leq a)$, that is, $a < b^*$, by Theorem 5.14(b). Since $b \in B$, we then have $b \leq a < b^*$. It follows from Theorem 5.14(c) that $a = b \in A \cap B$, and this element is the required least element of A.   □

EXERCISES

5.36 Use the pigeonhole principle to prove that N is not a finite set.

5.37 Prove the associative law for $+$ on N.

5.38 Show that $a + 1 = 1 + a \; \forall a \in N$.

5.39 Deduce the commutative law for $+$ on N from the previous exercise.

5.40 Prove the cancellation law for $+$ on N.

5.41 Prove that multiplication is well defined throughout N by M1 and M2.

5.42 Prove the distributive law for $\cdot$ over $+$ in N.

5.43 Prove the associative law for $\cdot$ in N.

5.44 Prove that $1a = a \ \forall a \in N$.

5.45 Prove that $ba + a = (b + 1)a \ \forall a, b \in N$.

5.46 Deduce the commutative law for $\cdot$ in N from the previous two exercises.

5.47 Prove that for $b, c \in N$, $c = bc \Rightarrow b = 1$.

5.48 Deduce the cancellation law for $\cdot$ in N from the previous exercise.

5.49 Prove that, for $a, b \in N$,

$$a < b \Rightarrow \forall c \in N \ a + c < b + c \Rightarrow \exists c \in N \ a + c < b + c \Rightarrow a < b.$$

5.50 Prove that, for $a, b \in N$,

$$a < b \Rightarrow \forall c \in N \ ac < bc \Rightarrow \exists c \in N \ ac < bc \Rightarrow a < b.$$

6
Cardinal Numbers

"The controversial topic of *Mengenlehre* (theory of sets, or classes,
particularly of infinite sets) created in 1874–95 by Georg Cantor
(1845–1915) may well be taken, out of its chronological order,
as the conclusion of the whole story."

E. T. Bell
Men of Mathematics

Anyone who can count has some idea of the meaning of the word "number", at
least in relation to the elements of a finite set. Thus, we have a clear conception
of what is meant by saying that the number of deadly sins, days of Christmas,
steps is 7, 12, 39 respectively. We have already capitalized on this intuition to
gain understanding of operations on sets. For example, Exercises 3.20. 3.23,
3.24 and Theorems 3.3, 5.1 respectively assert that if A, B are finite sets with
$|A| = a$, $|B| = b$, then

$$|A \mathbin{\dot\cup} B| = a + b, \tag{6.1}$$

$$|A \times B| = ab, \tag{6.2}$$

$$|A^n| = a^n, \tag{6.3}$$

$$|\mathcal{P}(A)| = 2^a, \tag{6.4}$$

$$|\mathrm{Map}(A, B)| = b^a. \tag{6.5}$$

It is the aim of this chapter to extend the idea of cardinality to infinite sets.
The result is an attractive theory that accords well with intuition, although
there are some surprises, pleasant and otherwise, along the way.

6.1 Cardinal Arithmetic

Let A and B be two arbitrary sets. Then we say that A and B are **equivalent** (or similar, or equipotent) if there is a bijection $\theta : A \to B$, and then we write $A \sim B$. It is easy to check that the formal properties of an equivalence relation hold for $\sim$, as follows.

(R) $A \sim A$ via the bijection 1_A.

(S) If $\theta : A \to B$ is a bijection, it has an inverse $\theta^{-1} : B \to A$ by Theorem 5.6. It is clear from the definition of inverse that $(\theta^{-1})^{-1} = \theta$, so θ^{-1} is also a bijection, again by Theorem 5.6.

(T) If $\theta : A \to B$, $\phi : B \to C$ are bijections, then so is the composite $\phi\theta : A \to C$, as it has inverse: $(\phi\theta)^{-1} = \theta^{-1}\phi^{-1}$.

It would be nice to be able to say at this point that $\sim$ is therefore an equivalence relation. Then we could define $|A|$ as the $\sim$-class of the set A, and cardinality as the natural map assigning $|A|$ to A, and we would have a *definition* of the concept of number in terms of the concept of set. It would then be possible to deduce properties of numbers, even infinite ones, from properties of sets, and so gain some understanding of the infinite. Well, this is more or less what happens. There is, however, a potentially dangerous obstacle in our path. We shall describe it now, and indicate how to get around it.

So why can't we say that equivalence of sets is an equivalence relation? It has the required properties (R), (S), (T) all right, and for any two sets A, B the set $\mathrm{Map}(A, B)$ either contains a bijection or it does not. The problem is that a relation has to be defined on a *set*, which in this case would appear to be the "set $\mathcal{S}$ of all sets". Now the "set" $\mathcal{S}$ would not only be truly enormous, but would also have the rather disturbing property of being a member of itself: $\mathcal{S} \in \mathcal{S}$. This is not only suspicious but altogether out of court. All right then, what about the "set $\mathcal{R}$ of all sets that are not members of themselves"? First check whether $\mathcal{R}$ is a member of itself or not:

$$\mathcal{R} \in \mathcal{R} \Rightarrow \mathcal{R} \text{ is not a member of itself: } \mathcal{R} \notin \mathcal{R}.$$

This contradiction proves that $\mathcal{R} \notin \mathcal{R}$, but then

$$\mathcal{R} \notin \mathcal{R} \Rightarrow \mathcal{R} \text{ is a member of itself: } \mathcal{R} \in \mathcal{R},$$

which is another contradiction. This rotten state of affairs is known as **Russell's paradox**. The consequence is that objects such as $\mathcal{R}$ and $\mathcal{S}$, if they exist at all, cannot reasonably be thought of as sets.

To get around this we shall have to make an assumption (or hypothesis or postulate). Such an hypothesis was framed by Cantor as follows.

Postulate 2

For every set A there exists an object $|A|$, called the **cardinal number** (or cardinality) of A, with the property that for any two sets A, B

$$|A| = |B| \Leftrightarrow \text{ there is a bijection} : A \to B.$$

This assertion can be justified by letting $|A|$ be the *smallest* set, in some sense, that is equivalent to A. The idea of ordering thus enters the picture, and there is a coherent theory of ordinal types and ordinal numbers whereby the word "smallest" can be given a precise meaning in this context. Since that theory is beyond our scope, we shall simply accept Cantor's postulate as a true statement and get back to work.

Thinking of cardinals as numbers, we would like to operate on them in a manner consistent with familiar usage in the finite case. Thus the **addition, multiplication** and **exponentiation** of cardinals are *defined* by formulae (6.1), (6.2) and (6.5) above. For example, to define the product of cardinals a and b, take sets A, with $|A| = a$, $|B| = b$ and put $ab = |A \times B|$. Similarly for the sum of a and b, where the condition of disjointness is easily assured (Exercise 6.1). Note that (6.3) and (6.4) are both special cases of (6.5). Finally, it has to be checked that these operations are well defined. This is a straightforward business and is consigned to the exercises.

Turning to the question of universal laws for cardinal arithmetic, some of these, such as the associative and commutative laws, are almost obvious. Others are consequences of bijections established in previous chapters. For example, Exercises 3.25 and 5.33–5.35 result in the following facts:

$$C \times (A \,\dot{\cup}\, B) \quad \sim \quad (C \times A) \,\dot{\cup}\, (C \times B), \tag{6.6}$$

$$\mathcal{P}(A \,\dot{\cup}\, B) \quad \sim \quad \mathcal{P}(A) \times \mathcal{P}(B), \tag{6.7}$$

$$\text{Map}(A \,\dot{\cup}\, B, C) \quad \sim \quad \text{Map}(A, C) \times \text{Map}(B, C), \tag{6.8}$$

$$\text{Map}(A \times B, C) \quad \sim \quad \text{Map}(A, \text{Map}(B, C)). \tag{6.9}$$

Letting $|A| = a$, $|B| = b$, $|C| = c$, we get the familiar distributive law and laws of indices:

$$c(a + b) \quad = \quad ca + cb, \tag{6.10}$$

$$2^{a+b} \quad = \quad 2^a 2^b, \tag{6.11}$$

$$c^{a+b} \quad = \quad c^a c^b, \tag{6.12}$$

$$c^{ab} \quad = \quad (c^b)^a, \tag{6.13}$$

where (6.11) is a special case of (6.12). All these laws are extensions of the corresponding ones for finite sets, and the latter can be accepted intuitively or deduced as in Section 5.4 from the Peano axioms. To complete this picture, it would be nice to extend the usual ordering $\leq$ from finite to infinite cardinals, and this is done as follows.

Given cardinal numbers $a = |A|$, $b = |B|$, we shall write $a \leq b$ if there is an injection $\alpha : A \to B$. To show that this coincides with the usual definition on finite cardinals is an easy exercise. To establish the formal conditions (R), (O), (T) for a partial ordering is, however, not quite so simple.

(R) Given $a = |A|$, the injection 1_A guarantees that $a \leq a$.

(O) The deduction of $a = b$ from $a \leq b$ and $b \leq a$ is not trivial and forms the subject of the next section.

(T) Given $a = |A|$, $b = |B|$, $c = |C|$ and injections

$$\alpha : A \to B, \qquad \beta : B \to C$$

the composite $\beta\alpha : A \to C$ is an injection, whence

$$a \leq b \leq c \Rightarrow a \leq c.$$

We conclude this section with an important and useful theorem. The proof, though somewhat redolent of Russell's paradox, is both elegant and valid. The idea is to compare, for any set A, the cardinals $|A|$ and $|\mathcal{P}(A)|$. First, it is clear that the mapping

$$\sigma : \quad A \quad \to \quad \mathcal{P}(A)$$
$$a \quad \mapsto \quad \{a\}$$

is an injection, so $|A| \leq \mathcal{P}(A)$, that is, $a \leq 2^a$ for any cardinal a. A potentially infinite sequence of infinite cardinals would then be guaranteed if we could prove that this equality is *strict*, and this is done as follows.

Theorem 6.1

For any set A, $|A| < |\mathcal{P}(A)|$.

Proof

Assume that $|A| = |\mathcal{P}(A)|$ and look for a contradiction. We thus have a bijection

$$\gamma : A \to \mathcal{P}(A)$$

assigning to any $a \in A$ a subset $\gamma(a) \subseteq A$. Now any $a \in A$ either belongs to its image $\gamma(a)$ or does not. So define the subset

$$B = \{a \in A \mid a \notin \gamma(a)\} \subseteq A.$$

Since $B \in \mathcal{P}(A)$ and γ is a surjection, there is an element $b \in A$ with $\gamma(b) = B$. Now ask the question: does b belong to B or not? Well,

$$b \in B \;\Rightarrow\; b \notin \gamma(b) \;\Rightarrow\; b \notin B, \text{ and}$$
$$b \notin B \;\Rightarrow\; b \in \gamma(b) \;\Rightarrow\; b \in B,$$

by the defining property of B in each case. This contradiction proves that no such bijection can exist. $\qquad\qquad\qquad\qquad\qquad\qquad\qquad\qquad\qquad\qquad\qquad\qquad$ $\square$

EXERCISES

6.1 Given arbitrary sets A and B, describe injections $\theta : A \to (A \cup B) \times \{0,1\}$, $\phi : B \to (A \cup B) \times \{0,1\}$ such that $\operatorname{Im}\theta \cap \operatorname{Im}\phi = \emptyset$.

6.2 Let A, S be sets with A arbitrary and $S = \{1,2,\ldots,n\}$, $n \in \mathbb{N}$. Prove that $A^n \sim \operatorname{Map}(S, A)$.

6.3 Prove that for any set A, $\mathcal{P}(A) \sim \operatorname{Map}(A, \{0,1\})$.

6.4 Let A, B and $\overline{A}, \overline{B}$ be sets with $A \cap B = \emptyset = \overline{A} \cap \overline{B}$. Prove that

$$A \sim \overline{A} \wedge B \sim \overline{B} \Rightarrow A \,\dot{\cup}\, B \sim \overline{A} \,\dot{\cup}\, \overline{B}.$$

6.5 For any sets $A, B, \overline{A}, \overline{B}$ prove that

$$A \sim \overline{A} \wedge B \sim \overline{B} \Rightarrow A \times B \sim \overline{A} \times \overline{B}.$$

6.6 For any sets $A, B, \overline{A}, \overline{B}$ prove that

$$A \sim \overline{A} \wedge B \sim \overline{B} \Rightarrow \operatorname{Map}(A, B) \sim \operatorname{Map}(\overline{A}, \overline{B}).$$

6.7 Convince yourself that addition and multiplication of cardinals satisfy the associative and commutative laws.

6.8 Let A and B be finite sets. Prove that there is an injection: $A \to B$ if and only if

$$|A| = |B| \text{ or } \exists n \in \mathbb{N} \quad |A| + n = |B|.$$

6.9 Consider the implication

$$\exists a \text{ surjection} : A \to B \Rightarrow |B| \leq |A|.$$

What must you assume to prove (a) this statement, (b) the converse?

6.10 Prove that, for any cardinals a and b,

$$b^a \leq 2^{ab}.$$

6.11 If a, b, c are cardinals and $a \leq b$, prove that $a^c \leq b^c$.

6.12 Given $a, b, c, d \in \mathbb{R}$ with $a < b$ and $c < d$, prove that the open intervals (a, b) and (c, d) are equivalent.

6.2 The Cantor–Schroeder–Bernstein theorem

This section is devoted to the deferred proof of property (O) for cardinal numbers. The statement is simple and the proof is elegant. The result is historically important and useful in practice. It thus has all the hallmarks of a top-class named theorem.

Theorem 6.2 (Cantor–Schroeder–Bernstein)

Let A, B be sets and

$$\theta : A \to B, \qquad \phi : B \to A$$

injections. Then $A \sim B$.

Proof

By regarding θ and ϕ as maps between $\mathcal{P}(A)$ and $\mathcal{P}(B)$, we shall construct a subset U of A such that ϕ maps $\theta(U)'$ surjectively onto U'. It will then follow from Exercise 6.4 that $A \sim B$.

First, extend the domain of θ to $\mathcal{P}(A)$ by setting, for $X \subseteq A$,

$$\theta(X) = \{\theta(x) \mid x \in X\} \subseteq B.$$

Similarly, for $X \subseteq B$ let

$$\phi(X) = \{\phi(x) \mid x \in X\} \subseteq A.$$

For the body of the proof we shall regard θ and ϕ in this way as maps between $\mathcal{P}(A)$ and $\mathcal{P}(B)$. Note that these maps preserve inclusions in an obvious sense. By contrast, the complementation maps

$$\alpha : \quad \mathcal{P}(A) \quad \to \quad \mathcal{P}(A) \qquad \beta : \quad \mathcal{P}(B) \quad \to \quad \mathcal{P}(B)$$
$$X \quad \mapsto \quad A \setminus X, \qquad\qquad X \quad \mapsto \quad B \setminus X$$

clearly reverse inclusions.

Next, and this is the crucial construction, form the composite map ω : $\mathcal{P}(A) \to \mathcal{P}(A)$ by:

(a) passing from $\mathcal{P}(A)$ to $\mathcal{P}(B)$ using θ,

(b) taking the complement in B,

(c) returning to $\mathcal{P}(A)$ via ϕ, and

(d) taking the complement in A.

Thus, ω is the map: $\mathcal{P}(A) \to \mathcal{P}(A)$ given by

$$\omega = \alpha\,\phi\,\beta\,\theta,$$

which, by earlier remarks, preserves inclusions.

Now consider the family

$$\mathcal{F} = \{X \in \mathcal{P}(A) \mid X \subseteq \omega(X)\}.$$

For example, $\emptyset \in \mathcal{F}$ always, but $A \in \mathcal{F}$ only when θ is surjective. All the members of $\mathcal{F}$ are subsets of A, and therefore so is their union:

$$U = \bigcup_{X \in \mathcal{F}} X \subseteq A.$$

Now watch carefully. For every $X \in \mathcal{F}$ we have

$$X \subseteq \omega(X) \text{ and } X \subseteq U.$$

Since ω preserves inclusions, it follows that

$$X \subseteq \omega(X) \subseteq \omega(U),$$

that is, every member of $\mathcal{F}$ is contained in $\omega(U)$ and so, therefore, is their union,

$$U \subseteq \omega(U).$$

Since ω preserves inclusions, we have

$$\omega(U) \subseteq \omega(\omega(U)),$$

that is, $\omega(U) \in \mathcal{F}$, and so $\omega(U) \subseteq U$, by definition of U. Thus, $U = \omega(U)$ and we have

$$\alpha(U) = \alpha(\omega(U)) = (\phi\,\beta\,\theta)(U),$$

since $\alpha^2 = 1_A$.

The final step in the proof is the interpretation of this formula. It says that $U' \subseteq A$ is the image under ϕ of $\theta(U)' \subseteq B$. So we have found partitions

$$A = U \,\dot{\cup}\, U', \qquad B = \theta(U) \,\dot{\cup}\, \theta(U)'$$

such that U is mapped bijectively to $\theta(U)$ by θ and $\theta(U)'$ is mapped bijectively to U' by ϕ. Then the map assigning to $x \in U$ its image under θ and to $x \notin U$ its pre-image under ϕ is the desired bijection from A to B. $\square$

This theorem will find application in the exercises below and in the next two sections in determining and comparing the cardinalities of various popular number systems and subsets of them.

EXERCISES

6.13 Prove that the map $\omega : \mathcal{P}(A) \to \mathcal{P}(A)$ in the proof of Theorem 6.2 preserves inclusions.

6.14 Show that $\emptyset \in \mathcal{F}$, but $A \notin \mathcal{F}$ when θ is not surjective.

6.15 Work through the proof of Theorem 6.2 in the special case when $A = B = \mathbb{N}$ and $\theta = \phi : n \mapsto n + 1$. In particular, prove that $\forall x \subseteq \mathbb{N} \; \omega(X) = \{x + 2 \mid x \in X\} \cup \{1\}$ and describe the set U.

6.16 In the special case when $A = B = \mathbb{N}$ and $\theta = \phi : n \mapsto 2n$, prove that $U = \{2^k l \mid k \text{ is even and } l \text{ is odd}\}$.

6.17 In the case when $A = B = \mathbb{N}$, $\theta(n) = 2n$, $\phi(n) = 3n$, characterize each of the sets U, U', $2U$, $(2U)'$ in terms of the highest powers of 2 and 3 dividing their elements.

6.18 Extend the result of Exercise 6.12 by proving that, for $a, b, c, d \in \mathbb{R}$ with $a < b$ and $c < d$, each of the intervals $(c, d]$, $[c, d)$, $[c, d]$ is equivalent to (a, b).

6.3 Countable Sets

We now know that any set S of cardinals is partially ordered by the relation $\leq$. A natural question to ask is whether this is a total ordering on S: given $a, b \in S$, can we assert that either $a \leq b$ or $b \leq a$? The answer is provided by the famous theorem of Zermelo, which asserts even more: any such S is *well ordered* by $\leq$. While the proof of this is beyond both our scope and needs, we shall derive the important special case (Theorem 6.3 below) that any cardinal is comparable with the cardinal $\aleph_0$ of the set $\mathbb{N}$ of positive integers. But first, a definition.

Definition 6.1

A set A is called **countable** if $|A| \leq \aleph_0$, that is, there is an injection $\alpha : A \to \mathbb{N}$.

In the special case when $\operatorname{Im} \alpha$ contains (a) the element 1, (b) no gaps, and (c) a maximal element, this corresponds precisely with our intuitive understanding of what it means to *count* the elements of A. Given any injection $\alpha : A \to \mathbb{N}$, properties (a) and (b) can be easily assumed by a process of "compression", so we assume them and concentrate on (c), which may or may not hold for our given α. These two cases are treated separately, as follows.

In the former case, when (c) holds, let a be maximal in $\operatorname{Im} \alpha$. Then $\operatorname{Im} \alpha = \{1, 2, \ldots, a\}$ and it makes sense to say that $|A| = a$. Then there can be no injection from $\mathbb{N}$ to A (see Section 5.3), and we can write $|A| < \aleph_0$. In this case, which includes the trivial case when $A = \emptyset$ ($a = 0$), we say that the set A and its cardinal a are **finite**.

Now suppose that (c) fails, so that $\operatorname{Im} \alpha$ has no maximal element Then $\operatorname{Im} \alpha$ satisfies the requirements of Peano's axiom P3 and we deduce that $\operatorname{Im} \alpha = \mathbb{N}$. Then α is a bijection and $|A| = \aleph_0$. In this case, we say that the set A and its cardinal $|A| = a$ are **countably infinite**. (In some books, the term "countable" refers only to this case.)

Finally, let A be an **uncountable** set, that is, a set for which there is no injection into $\mathbb{N}$. Then $A \neq \emptyset$ and $\exists x \in A$. Now consider the proposition:

$$P(n) \qquad \forall n \in \mathbb{N} \ \exists A_n \subseteq A \qquad |A_n| = n.$$

Then $P(1)$ is true: take $A_1 = \{x_1\}$ where $x_1 = x$. Now let $n > 1$ and assume $P(n-1)$:

$$\text{IH} \qquad \exists A_{n-1} = \{x_1, \ldots, x_{n-1}\} \subseteq A.$$

Since there is no injection from A into $\mathbb{N}$, it follows that $A_{n-1} \neq A$, that is, $A \setminus A_{n-1} \neq \emptyset$. Then pick $x_n \in A \setminus A_{n-1}$ and put $A_n = A_{n-1} \cup \{x_n\}$. Then $|A_n| = n$ and $P(n)$ is established. By the PMI, $P(n)$ is true $\forall n \in \mathbb{N}$. We have thus shown that $n \leq |A|$ $\forall n \in \mathbb{N}$. But the above construction actually proves something more. Since the sets A_n form a *chain*, $A_n \subseteq A_{n+1}$ $\forall n \in \mathbb{N}$, their union $U = \bigcup_{n \in \mathbb{N}} A_n$ is a set $\{x_n \mid n \in \mathbb{N}\}$ of cardinality $\aleph_0$. It follows that $\aleph_0 < |A|$ in this case, which completes the proof of the following theorem.

Theorem 6.3

Any cardinal a satisfies exactly one of the following three conditions:

$$a < \aleph_0, \qquad a = \aleph_0, \qquad a > \aleph_0,$$

that is, a is finite, countably infinite, or uncountable. $\qquad\qquad \square$

We know from the definition that among the cardinalities of countable sets $\aleph_0$ is the largest. From this theorem it follows that among the cardinalities of infinite sets, $\aleph_0$ is the smallest. We will come across some bigger ones in the next section.

In the rest of this section, we shall discuss properties of countable cardinals and, in particular, what happens when we start adding and multiplying them together. The main emphasis will be on countably infinite sets, and cases involving finite sets will largely be left to the exercises. We shall prove three major theorems. The first of these has the consequence that

$$|\mathbb{Z}| = \aleph_0. \tag{6.14}$$

Theorem 6.4

The union of two disjoint countable sets is countable. In particular,

$$\aleph_0 + \aleph_0 = \aleph_0. \tag{6.15}$$

Proof

Given disjoint sets A and B and injections

$$\theta : A \longrightarrow \mathbb{N}, \qquad \phi : B \longrightarrow \mathbb{N},$$

consider the map

$$
\begin{aligned}
\psi : \quad A \,\dot\cup\, B \quad &\longrightarrow \quad \mathbb{N}, \\
x \quad &\longmapsto \quad
\begin{cases}
2\theta(x), & x \in A, \\
2\phi(x) + 1, & x \in B.
\end{cases}
\end{aligned}
$$

Since θ is an injection, so is $\psi|_A$, for

$$2\theta(x) = 2\theta(y) \Rightarrow \theta(x) = \theta(y) \Rightarrow x = y.$$

Similarly, $\psi|_B$ is an injection. Moreover, the values of $\psi|_A$ are all even and the values of $\psi|_B$ are all odd, so that $\operatorname{Im}\psi|_A \cap \operatorname{Im}\psi|_B = \emptyset$. Thus, ψ is an injection.

In the special case when A and B are disjoint countably infinite sets, we deduce that $|A \,\dot\cup\, B| \leq \aleph_0$. Since $A \,\dot\cup\, B$ is obviously not finite, it follows from Theorem 6.3 that $|A \,\dot\cup\, B| = \aleph_0$, which proves (6.15). $\qquad\qquad\square$

We now focus attention on the Cartesian product of two countably infinite sets. Since the multiplication of cardinals is well defined (Exercise 6.5), it is sufficient to consider the set $\mathbb{N} \times \mathbb{N} = \mathbb{N}^2$. The next theorem has the consequence that

$$|\mathbb{Q}| = \aleph_0. \qquad\qquad (6.16)$$

Theorem 6.5

The Cartesian product of two countably infinite sets is countably infinite, that is,

$$\aleph_0 \cdot \aleph_0 = \aleph_0. \qquad\qquad (6.17)$$

Proof

We have to count pairs (k, l) of positive integers, and we do this in accordance with the lexicographic ordering of the corresponding pairs $(k + l, l)$. This is illustrated in Fig. 6.5, where the integer assigned to (k, l) occupies the usual position of (k, l) in the positive quadrant of the Cartesian plane.

```
15
10   14
 6    9   13
 3    5    8   12      ·.
 1    2    4    7   11   16
```

Fig. 6.1. Counting the elements of $\mathbb{N} \times \mathbb{N}$.

To be more specific, let

$$A_n = \{(k, l) \mid k + l = n\},$$

$n = 2, 3, 4, \ldots$, so that $|A_n| = n - 1$. Next put

$$U_n = \bigcup_{k=2}^{n} A_k,$$

so that $|U_n| = \frac{1}{2}(n-1)n = u_n$, say, $n \geq 2$. The integer thus assigned to the pair (k, l), where $k + l = n$, is then given by

$$\alpha(k, l) = u_{n-1} + l. \tag{6.18}$$

This defines a map $\alpha : \mathbb{N}^2 \to \mathbb{N}$, which we claim is an injection.

To prove this, let (k, l), $(k', l') \in \mathbb{N}^2$ and assume that $\alpha(k, l) = \alpha(k', l')$, that is,

$$u_{n-1} + l = u_{m-1} + l', \tag{6.19}$$

where $n = k + l$ and $m = k' + l'$. There are two cases to consider. If $m = n$, then $l = l'$ by (6.19), and so $k = n - l = m - l' = k'$, whence $(k, l) = (k', l')$ as required. If $m \neq n$, we may assume by symmetry that $m < n$. Then, since $l' < m$,

$$u_{n-1} + l = u_{m-1} + l' < u_{m-1} + m = u_m + 1 \leq u_{n-1} + 1 \leq u_{n-1} + l.$$

This contradiction shows that the case $m \neq n$ cannot arise. The assumption (6.19) thus forces $(k, l) = (k', l')$, and α is an injection, as claimed.

We have shown that $\aleph_0^2 \leq \aleph_0$. Since $\mathbb{N}^2$ is obviously not a finite set, it follows from Theorem 6.3 that $\aleph_0^2 = \aleph_0$. $\qquad\square$

The final theorem in this section is used often and widely in pure mathematics. We prove it now as a consequence of Theorem 6.5.

Theorem 6.6

A countable union of countable sets is countable.

Proof

Let the sets be A_i, $i \in I$, where each A_i is countable and so is I. Thus there are injections

$$\theta_i : A_i \to \mathbb{N}, \ i \in I, \ \text{and} \ \phi : I \to \mathbb{N}.$$

We obtain pairwise disjoint copies of the A_i by putting

$$\overline{A_i} = \{(\theta_i(a), \ \phi(i)) \mid a \in A_i\} \subseteq \mathbb{N}^2$$

for each $i \in I$. These are copies (in the sense that $\overline{A_i} \sim A_i \ \forall i \in I$) since every θ_i is an injection, and pairwise disjoint because ϕ is an injection.

We will now define a map

$$\sigma : \overline{A} = \bigcup_{i \in I} \overline{A_i} \longrightarrow A = \bigcup_i A_i.$$

A typical element $\overline{a} \in \overline{A}$ has the form $\overline{a} = (\theta_i(a), \phi(i))$ for some $i \in I$, $a \in A_i$. Then put $\sigma(\overline{a}) = a \in A_i \subseteq A$. Since every $a \in A$ belongs to A_i for some $i \in I$, we have $a = \sigma((\theta_i(a), \phi(i)))$, and σ is a surjection. By the Axiom of Choice, the partition $\{\sigma^{-1}(a) \mid a \in A\}$ of $\overline{A}$ has a transversal T. Then the map $\tau : A \to \overline{A}$ with $\tau(a) = \overline{a}$, where $T \cap \sigma^{-1}(a) = \{\overline{a}\}$, has the property $\sigma\tau = 1_A$, and τ is thus an injection from A to a subset $\overline{A}$ of $\mathbb{N}^2$. It follows from the transitivity of $\leq$ that $|A| \leq \aleph_0^2$, and then by the previous theorem that A is countable, as required. $\qquad \square$

EXERCISES

6.19 Let $(S, \sigma, 1)$, $(T, \tau, 1)$ be two Peano triples, that is, S and T are sets with $1 \in S$ and $1 \in T$ and $\sigma : S \to S$, $\tau : T \to T$ are mappings such that the three Peano axioms hold. Prove that $|S| = |T|$ by constructing a bijection $\beta : S \to T$.

6.20 Let a and b be cardinals with $a + 1 = b + 1$. Prove that $a = b$.

6.21 If a is a cardinal, prove that $a < a + 1 \Leftrightarrow a$ is finite.

6.22 Prove that a set A is infinite if and only if $\exists B \subset A$ with $B \sim A$.

6.23 Let a and b be cardinals with a finite and b infinite. Prove that $a < b$.

6.24 Prove that $\aleph_0 + n = \aleph_0 \ \forall n \in \mathbb{N}$.

6.25 Prove that $n\aleph_0 = \aleph_0 \ \forall n \in \mathbb{N}$.

6.26 Prove that $\aleph_0^n = \aleph_0 \ \forall n \in \mathbb{N}$.

6.27 Prove that the Cartesian product of two countable sets is countable.

6.28 Use Theorem 6.4 to prove that $|\mathbb{Z}| = \aleph_0$.

6.29 Use Theorem 6.5 to prove that $|\mathbb{Q}| = \aleph_0$.

6.4 Uncountable Sets

The existence of uncountable sets has already been foreshadowed in Theorem 6.1: the power set $\mathcal{P}(\mathbb{N})$ is uncountable. Most of this section will be devoted to the study of this enormous set.

We begin by recalling (Exercise 5.20) one method of calculating the cardinality of $\mathcal{P}(S)$ when S is a finite set. Let $S = \{k \in \mathbb{N} \mid 1 \leq k \leq n\}$, $n \in \mathbb{N}$, and suppose $A \subseteq S$. Then we define the characteristic function of A as follows:

$$\chi_A: \quad \begin{aligned} S &\longrightarrow \mathbb{Z}/2\mathbb{Z} = \{0,1\} \\ x &\longmapsto \begin{cases} 1 & \text{if } x \in A, \\ 0 & \text{if } x \notin A. \end{cases} \end{aligned} \tag{6.20}$$

Since the map $\chi: \mathcal{P}(S) \to \operatorname{Map}(S, \mathbb{Z}/2\mathbb{Z})$, $A \mapsto \chi_A$, is a bijection (easy exercise), we deduce that $|(\mathcal{P}(S))| = 2^n$. In exactly the same way, we get a bijection

$$\chi: \quad \begin{aligned} \mathcal{P}(\mathbb{N}) &\longrightarrow \operatorname{Map}(\mathbb{N}, \mathbb{Z}/2\mathbb{Z}) \\ A &\longmapsto \chi_A, \end{aligned} \tag{6.21}$$

with $\chi_A(X)$ defined as in (6.20).

The map χ_A is traditionally thought of as a sequence, or "infinite string", of zeros and ones, namely an expression of the form

$$\rho_A = .a_1 a_2 \ldots a_n \ldots, \tag{6.22}$$

where $a_n = \chi_A(n)$, $n \in \mathbb{N}$, and the "." at the start is called the **binary point**. For example, if $A = E = 2\mathbb{N}$, the set of even numbers, then

$$\rho_E = .0101 \ldots a_n \ldots, \tag{6.23}$$

where a_n is 1 if n is even and 0 if n is odd. Similarly, if $A = O = 1 + 2\mathbb{N}$, the set of odd numbers, then

$$\rho_O = .1010 \ldots a_n \ldots, \tag{6.24}$$

where a_n is 0 if n is even and 1 if n is odd. The alternating strings (6.23), (6.24) are often abbreviated to $.0\dot{1}$, $.1\dot{0}$ respectively, and are examples of recurring strings.

Similar examples of recurring strings are as follows. Suppose that, beyond some point, $n \geq m$ say, all the a_n are equal, to $a \in \{0,1\}$ say. Then, depending on the value of a, we get a recurring string

$$\rho_A = .a_1 \ldots a_m \dot{0} \text{ or } .a_1 \ldots a_m \dot{1}. \tag{6.25}$$

These strings clearly correspond to the cases where the set A is finite or cofinite (has finite complement) respectively.

Returning to the general case, we now associate to the string (6.22) a *sequence of rational numbers*, that is, a map $\beta_A : \mathbb{N} \to \mathbb{Q}$, as follows:

$$\text{for } n \in \mathbb{N}, \quad \beta_A(n) = \sum_{k=1}^{n} a_k 2^{-k}. \tag{6.26}$$

Next, observe that when $m \geq n \geq N$, for some fixed $N \in \mathbb{N}$,

$$\begin{aligned}
\beta_A(m) - \beta_A(n) &= \sum_{k=n+1}^{m} a_k 2^{-k} \leq \sum_{k=n+1}^{m} 2^{-k} \\
&= 2^{-n} - 2^{-m} < 2^{-n} \leq 2^{-N}.
\end{aligned}$$

This shows that the terms of the sequence (6.26) are arbitrarily close eventually, in precisely the sense of our definition of *Cauchy sequence* in Chapter 4. Viewed in this light, formula (6.22) is nothing but the binary expansion of the **real number** ρ_A.

We are now very close to a bijection between subsets A of $\mathbb{N}$ and real numbers ρ_A in the interval $[0, 1]$. There is one problem remaining, caused by the expansions $.a_1 \ldots a_n \dot{1}$ corresponding to the cofinite sets. For example, take the number $\rho_A = .0\dot{1}$ corresponding to the subset $A = \mathbb{N} \setminus \{1\}$. According to (6.26), the terms of the corresponding Cauchy sequence are

$$\sum_{k=2}^{n} 2^{-k} = \frac{1}{2} - \left(\frac{1}{2}\right)^n,$$

which differ from the constant sequence $\frac{1}{2}$ by the null sequence $(\frac{1}{2})^n$. The constant sequence $\frac{1}{2}$ corresponds to the binary expansion $.1\dot{0}$ of the singleton $\{1\}$, and so the distinct subsets $\{1\}$ and $\mathbb{N} \setminus \{1\}$ have equivalent Cauchy sequences, that is, they correspond to the same real number. The same argument shows that the distinct strings

$$.a_1 \ldots a_n 0\dot{1}, \qquad .a_1 \ldots a_n 1\dot{0} \tag{6.27}$$

correspond to the same real number.

It turns out, however, that the *only* violations of injectivity are those of the form (6.27), and we shall prove this now. The claim is that two distinct strings

$$a = .a_1 \ldots a_n \ldots, \qquad b = .b_1 \ldots b_n \ldots$$

correspond to distinct real numbers provided that neither terminates with an infinite string $\dot{1}$ of ones. Assume this proviso, and let the first place where a and b differ be the nth:

$$a_k = b_k, \quad 1 \leq k \leq n-1, \quad a_n = 1, \quad b_n = 0,$$

say. By hypothesis, there is at least one 0 to the right of b_n in b; let the first of these be b_m:

$$m > n, \quad b_m = 0, \quad b_k = 1 \quad \text{for} \quad n < k < m.$$

Then the terms of the Cauchy sequence for a are eventually greater than or equal to those of the constant sequence for

$$a' = .a_1 \ldots a_{n-1}1\dot{0},$$

and those of the Cauchy sequence for b are eventually less than or equal to those of the constant sequence for

$$b' = .a_1 \ldots a_{n-1}01 \ldots 1\dot{0},$$

where the final one is in the mth place. Since $b' < a'$, it follows that the Cauchy sequences for a and b cannot differ by a null sequence, as their terms eventually differ by at least $a' - b' = 2^{-m}$.

We summarize what we have proved so far in the form of a theorem.

Theorem 6.7

There is a bijection between the non-cofinite members of $\mathcal{P}(\mathbb{N})$ and the real numbers in the interval $[0, 1)$. □

The interval here is half-open as we are allowing $0 = .\dot{0}$, which corresponds to $\emptyset$, but not $1 = .\dot{1}$, which corresponds to the cofinite set $\mathbb{N}$.

Our next job is to count the cofinite subsets of $\mathbb{N}$, or, equivalently, the number of finite subsets (complementation is a bijection between these families). So how many subsets of $\mathbb{N}$ are there with n elements? Certainly no more than $|\operatorname{Map}(A, \mathbb{N})|$, where $|A| = n$, that is, at most $\aleph_0^n = \aleph_0 \; \forall n \in \mathbb{N}$ (by Exercise 6.26). Thus, the number of n-element subsets of $\mathbb{N}$ is countable $\forall n \geq 0$. Since the set of $n \geq 0$ is countable, our next result follows from Theorem 6.6.

Theorem 6.8

The number of cofinite subsets of $\mathbb{N}$ is $\aleph_0$. □

There is one more link in the chain connecting the power set $\mathcal{P}(\mathbb{N})$ and the set $\mathbb{R}$ of all real numbers. Before forging it, here is an important definition.

Definition 6.2

The cardinality of the set $\mathbb{R}$ of all real numbers is called the **cardinal of the continuum** and denoted by c.

For our last step, put

$$\mathbb{R}^+ = \{r \in \mathbb{R} \mid r > 0\}$$

and consider the map

$$\alpha: \quad (0,1) \longrightarrow \mathbb{R}^+ \atop x \longmapsto \frac{x}{1-x}. \tag{6.28}$$

Since α has an inverse,

$$\beta: \quad \mathbb{R}^+ \longrightarrow (0,1) \atop r \longmapsto \frac{r}{1+r},$$

α is a bijection. By composing α with the obvious bijections from $\mathbb{R}^+$ to $\mathbb{R}^- = \{r \in \mathbb{R} \mid r < 0\}$ and from $(-1,0)$ to $(1,0)$, we get a bijection from $(-1,0)$ to $\mathbb{R}^-$. Combining this map with α (and $0 \longmapsto 0$), we get a bijection from $(-1,1)$ to $\mathbb{R}$. Since all intervals with endpoints a and b, $a < b$, have the same cardinality as $(-1,1)$, we have proved the following result.

Theorem 6.9

The cardinality of any interval with endpoints a and b, $a < b$, is c.

We are now in a position to prove our last theorem.

Theorem 6.10

$\mathcal{P}(\mathbb{N})$ has the cardinal c of the continuum.

Proof

By Theorems 6.7 and 6.8,

$$|\mathcal{P}(\mathbb{N})| = |[0,1)| + \aleph_0.$$

Since $\aleph_0 \leq c$ (Theorem 6.3), it follows from Theorem 6.9 that

$$|\mathcal{P}(\mathbb{N})| \leq |[0,1)| + |[1,2]| = |[0,2]|.$$

So Theorems 6.7 and 6.9 together yield

$$c \leq |\mathcal{P}(\mathbb{N})| \leq c,$$

and the proof is completed by an application of the Cantor–Schroeder–Bernstein theorem.   □

We thus have a fairly well-understood cardinal c greater than $\aleph_0$. There are even bigger ones of course, such as $|\mathcal{P}(\mathbb{R})|$. An obvious question is: are there any that lie strictly between $\aleph_0$ and c? That the answer to this question is in the negative forms the famous **continuum hypothesis** of Cantor, which occupied logicians for many decades of this century. It was finally proved to be independent of the usual axioms of set theory by P. J. Cohen in 1963.

EXERCISES

6.30 Describe the inverse of the map $\chi : \mathcal{P}(\mathbb{N}) \to \mathrm{Map}(\mathbb{N}, \mathbb{Z}/2\mathbb{Z})$.

6.31 What real numbers are represented by the binary expansions corresponding to the subsets O and E of $\mathbb{N}$?

6.32 Prove that recurring binary expansions correspond exactly to the rational numbers in the interval $[0, 1]$.

6.33 Use the previous exercise to give an alternative proof of Theorem 6.8.

6.34 Describe explicitly the bijection from $(-1, 0)$ to $\mathbb{R}^-$ obtained from (6.28).

6.35 How might Theorem 6.7 be used to compare ($\leq$), add ($+$) and multiply ($\cdot$) two elements of $[0, 1)$? How might these definitions be extended to the whole of $\mathbb{R}$?

6.36 Use a suitable function: $\mathbb{R} \to \mathbb{R}$ to prove that $2c + 1 = c$ (cf. Exercise 6.28).

6.37 Prove that, for any cardinal $b \leq c$, $b + c = c$.

6.38 By "intertwining" the binary expansions of two numbers in $(0, 1)$, prove that $c^2 = c$.

Solutions to Exercises

Chapter 1

1.3 By the generalized associative law brackets are not needed, and then the ordinary commutative law, used $\frac{1}{2}n(n-1)$ times, does the rest.

1.4 The variable of summation, k or i, is a so-called "dummy variable": the value of the sum does not depend on the symbol used to represent it.

1.6 Replace k by $n - k$ throughout the right-hand side. The limits are $n - k = 0$ and $n - k = n - 1$, that is, $k = n$ and $k = 1$.

1.8 In the sum in parentheses on the right-hand side, $a = a_i$ does not depend on the variable j of summation, so we can pull it out to the left by Exercise 1.2. The sum $\sum_{j=1}^{m} b_j$ being independent of i, we can similarly pull it out to the right.

1.9 Both sides are equal to the sum of the $a_i b_j$ over all integer values of i and j in the range $0 \leq i < j \leq n$

1.11 By Exercise 1.6, $s_n = \sum_{k=0}^{n-1} a_k = \sum_{k=1}^{n} a_{n-k}$. By Exercise 1.5 we get $s_n = \sum_{k=1}^{n} a_{k-1}$. Thus,

$$2s_n = \sum_{k=1}^{n} a_{k-1} + \sum_{k=1}^{n} a_{n-k}$$

$$= \sum_{k=1}^{n} (a_{k-1} + a_{n-k}), \text{ by Exercise 1.3,}$$

$$= \sum_{k=1}^{n}(2a + (n-1)d), \text{ by } (1.5),$$

$$= n(2a + (n-1)d), \text{ by Exercise } 1.1.$$

Dividing both sides by 2 and applying the distributive law, we get (1.7).

1.12 Take $n = 4$ and label the three additions $1, 2, 3$ in the order they are carried out. Then the $3! = 6$ permutations

$$123, \quad 132, \quad 213, \quad 231, \quad 312, \quad 321$$

correspond to the bracketings

$$((\cdot\cdot)\cdot)\cdot, \quad (\cdot\cdot)(\cdot\cdot), \quad (\cdot(\cdot\cdot))\cdot, \quad \cdot((\cdot\cdot)\cdot), \quad (\cdot\cdot)(\cdot\cdot), \quad \cdot(\cdot(\cdot\cdot)),$$

and two of these, the second and fifth, are the same. Hence, $c_4 = 5$. Similarly, $c_5 = 14$, $c_6 = 42$, and so on. If the ith addition is performed last, we get $c_i c_{n-i}$ bracketings, whence

$$c_n = \sum_{i=1}^{n-1} c_i c_{n-i}.$$

A closed formula for c_n appears in Section 1.5.

1.13 We must deduce (1.14) from (1.15). So assume (1.15) and let $a \neq 0$ and $b \neq 0$. Then a^{-1} and b^{-1} exist, and

$$ab \cdot b^{-1}a^{-1} = 1,$$

so that ab cannot be zero.

1.14 If an integral domain is finite, we can list its elements a_i, $1 \leq i \leq n$. If a is one of these and $a \neq 0$, we claim that the products aa_i, $1 \leq i \leq n$, are all different. For if $aa_i = aa_j$, then $a(a_i - a_j) = 0$, and (1.15) implies that $a_i = a_j$, whence $i = j$. Thus (pigeon-hole principle, see Chapter 5), some $aa_k = 1$. Then $a_k = a^{-1}$, whence (1.15) holds and we have a field.

1.15 $1/2, 0/1, -2/3, 6/1, 1/2, -2/3$. The second and fourth of these would normally be further simplified to $0, 6$, respectively.

1.16 Suppose $r = a/b = c/d$, where b and d both satisfy (1.17). Then $b = d$ as they are both positive and minimal. Multiplying the equation $a/b = c/b$ by b, we get $a = c$, and the expressions a/b and c/d are identical.

1.18 Assume for a contradiction that c is rational: $c = a/b$ in lowest terms. Then $a^3 = 5b^3$, whence a^3 and also a is divisible by 5, $a = 5c$ say. Then $125c^3 = a^3 = 5b^3$, so that $b^3 = 25c^3$, whence b^3 and also b is divisible by 5, contradicting the fact that a/b is in lowest terms. So c is irrational.

1.19 The negation of "P is false" is "P is not false", which is the same as "P is true".

Assume P and deduce $\sim P$. We now have P and $\sim P$, which is a contradiction. We conclude that P is false, that is, (b). We can also conclude (c) as it is the same as (b). (a) and (d) are also the same, and cannot be concluded.

1.20 Assume for a contradiction that P is false, and let $n - 1$, n, $n + 1$ be integers for which it fails. Thus, there is an integer m such that

$$12m - 1 = (n - 1)^2 + n^2 + (n + 1)^2,$$

that is,

$$12m = 3(n^2 + 1).$$

Thus, $n^2 + 1 = 4m$, so that n^2 leaves remainder -1 on division by 4, which contradicts the first part of Exercise 1.17. So P is true.

1.22 Assume that $f(x) = g(x) h(x)$, where $g(x)$ and $h(x)$ are quadratics over $\mathbb{Z}$. Comparing coefficients of x^4, the leading coefficients of $g(x)$ and $h(x)$ are both equal to 1 or -1. In the latter case, replace $g(x)$ and $h(x)$ by their negatives. Then we can write

$$g(x) = x^2 + ax + b, \quad h(x) = x^2 + cx + d,$$

where a, b, c, d are integers. Multiplying out and comparing coefficients, we get

$$a + c = 0, \quad ac + b + d = 2, \quad ad + bc = 2, \quad bd = 1998.$$

Use the first two equations to eliminate $c = -a$, $d = a^2 - b + 2$ from the third: $a(a^2 - 2b + 2) = 2$. Thus, a is even and 2 is the product of two even numbers, which is impossible.

1.24 Let the people be $1, 2, \ldots, n$, with $f_1, f_2, \ldots, f_n$ friends present, respectively. Assume for a contradiction that the numbers $f_1, f_2, \ldots, f_n$ are all different. Since each of the numbers is between 0 and $n-1$ inclusive, they must be equal to $0, 1, \ldots, n - 1$ in some order. So we can find i and j such that $f_i = 0$ and $f_j = n - 1$. So j is everyone's friend, in particular, i and j are friends, and i is no-one's friend, in particular, i and j are not friends. Contradiction.

1.25 $b = a$, $b < a$, $b \mid a$, b and a have no common prime divisor.

1.26 $a \neq b$, $a \geq b$, $a \nmid b$, there is a prime p such that $p \mid a$ and $p \mid b$.

1.28 When $a = 2$, $b = 1$, B' is true but $\sim D$ is false.

When $a = 1$, $b = 2$, D' is true but $\sim B$ is false. In the remaining nine cases, $a = b = 1$ does the trick.

1.29 If a positive integer $n \geq 2$ is prime, then it is divisible by no integer d in the range $2 \leq d \leq \sqrt{n}$. Yes.

1.30 It was not correct: given that $A \Rightarrow B$ and B, you cannot deduce A. In fact, they both slipped on some rabbit droppings.

1.32 Assume the negation of the conclusion: $\sqrt{ab} = (a + b)/2$. Then

$$
\begin{aligned}
ab &= (a + b)^2/4 = (a^2 + 2ab + b^2)/4 \\
   &= (a^2 - 2ab + b^2)/4 + ab \\
   &= (a - b)^2/4 + ab.
\end{aligned}
$$

Thus, $(a - b)^2 = 0$ and so $a = b$, which is the negation of the premise.

1.34 Both sides are equal to 1 when $n = 1$.

Let $n > 1$ and assume the

$$
\text{IH:} \qquad \sum_{k=1}^{n-1} k^3 = \frac{1}{4}(n-1)^2 n^2
$$

$$
\begin{aligned}
\text{Then} \qquad \sum_{k=1}^{n} k^3 &= \frac{1}{4}(n-1)^2 n^2 + n^3 \\
&= \frac{1}{4} n^2 ((n-1)^2 + 4n) \\
&= \frac{1}{4} n^2 (n+1)^2.
\end{aligned}
$$

For the rest, consider an $n \times n$ square of numbers with lm in the l-row and m-column. Check that

(i) the sum of the n terms in the l-row is $l \sum_{k=1}^{n} k$, and

(ii) the sum of the first k terms of the k-row and the first $k - 1$ terms of the k-column is k^3.

1.35
$$s_4(x) = 4\int \tfrac{1}{4}(x^4 + 2x^3 + x^2)\,dx + cx$$

$$= \tfrac{1}{5}x^5 + \tfrac{1}{2}x^4 + \tfrac{1}{3}x^3 - \tfrac{1}{30}x.$$

$P(n):$
$$\sum_{k=1}^{n} k^4 = \tfrac{1}{30}(6n^5 + 15n^4 + 10n^3 - n).$$

When $n = 1$, both sides are equal to 1.

Let $n > 1$ and assume $P(n-1)$. Then

$$\sum_{k=1}^{n} k^4 = \frac{1}{30}(6(n-1)^5 + 15(n-1)^4 + 10(n-1)^3 - n) + n^4$$

$$= \frac{1}{30}(6n^5 + 15n^4 + 10n^3 - n), \text{ after some fiddling about,}$$

as required.

1.36 The inductive base is sound and so is the inductive step for $n \geq 3$, but it fails for $n = 2$. Moral: keep a close check on the value of the inductive variable.

1.37 $\sum_{k=0}^{n-1} ar^k = a\sum_{k=0}^{n-1} r^k = a(r^n - 1)/(r - 1)$.

When $n = 1$, both sides are equal to a. Let $n \geq 1$ and assume the formula for n. Then

$$\sum_{k=0}^{n} ar^k = a\frac{r^n - 1}{r - 1} + ar^n$$

$$= a\frac{r^n - 1 + r^{n+1} - r^n}{r - 1}$$

$$= a\frac{r^{n+1} - 1}{r - 1}.$$

1.39 Suppose for a contradiction that there are only finitely many, say $p_1, p_2, \ldots, p_n$. By Theorem 1.7, the number $N = p_1 p_2 \cdots p_n + 1$ is a product of primes. In particular N is divisible by some prime, p say. Then p is equal to some p_i, $1 \leq i \leq n$. Since $p \mid N$ and $p \mid p_1 \cdots p_n$, we deduce that $p \mid (N - p_1 \cdots p_n) = 1$. Contradiction.

1.42 There are several possible answers here, one of which is as follows:

prove $P(1,1)$, assume $P(1, n-1)$, $n > 1$, prove $P(1,n)$, assume $P(m-1,1)$, $m > 1$, prove $P(n,1)$, assume both $P(m-1,n)$, and $P(m, n-1)$, $m, n > 1$, prove $P(m,n)$.

1.43 Follow the strategy of the previous exercise or that described in the text. Take the latter case and $P(m,n) : m + n = n + m$. $P(1,1)$ is obvious.

Let $m \geq 1$ and assume $P(m,1) : m + 1 = 1 + m$.

Prove $P(m+1,1)$:

$$
\begin{aligned}
(m+1) + 1 &= (1+m) + 1 \quad \text{by the IH} \\
&= 1 + (m+1) \quad \text{by the associative law.}
\end{aligned}
$$

Let $n \geq 1$ and assume $P(m,n)$, $m + n = n + m$ for all m.

Prove $P(m, n+1)$ for all m:

$$
\begin{aligned}
m + (n+1) &= (m+n) + 1 \quad \text{by the associative law} \\
&= (n+m) + 1 \quad \text{by the IH} \\
&= n + (m+1) \quad \text{by the associative law} \\
&= n + (1+m) \quad \text{by } P(m,1) \\
&= (n+1) + m \quad \text{by the associative law,}
\end{aligned}
$$

as required.

1.44 $\pi(m,n)/n!$ is just another way of writing the binomial coefficient $\binom{m+n}{n}$, which is an integer.

1.45 $\sigma_1 = a_1$, $\sigma_n = \sigma_{n-1} + a_n$ for all $n \geq 2$. $\pi_1 = a_1$, $\pi_n = \pi_{n-1} \cdot a_n$ for all $n \geq 2$.

1.46 When $m = 0$, both sides of the equation $(x^{-1})^m = (x^m)^{-1}$ are equal to 1. When m is positive

$$
x^m x^{-m} = x^m (1/x)^m = (x/x)^m = 1,
$$

whence

$$
(x^{-1})^m = x^{-m} = 1/x^m = (x^m)^{-1}.
$$

Now let m be negative, say $m = -n$ with $n \geq 1$. Then $(x^{-1})^m = (x^{-1})^{-n} = ((x^{-1})^{-1})^n = ((x^{-1})^n)^{-1} = (x^{-n})^{-1} = (x^m)^{-1}$, where the positive case is used at the third step. Thus

$$
(x^{-1})^m = (x^m)^{-1} \text{ for all integers } m, \tag{$*$}
$$

as required.

To get the rules of indices in the general case, first observe that the proof of Theorem 1.9 works for all integer values of m (check) provided $n \geq 0$. Replacing x by x^{-1} in the first rule, with m arbitrary and $n \geq 0$,

$$(x^{-1})^m \cdot (x^{-1})^n = (x^{-1})^{m+n},$$

that is,

$$x^{-m}x^{-n} = x^{-m-n}$$

for all m and $n \geq 0$, whence (replacing m by $-m$)

$$x^m x^{-n} = x^{m-n},$$

which proves the first rule. For the second, let $n > 0$ and observe that, using $(*)$,

$$(x^m)^{-n} = ((x^m)^{-1})^n = (x^{-m})^n = x^{-mn}.$$

1.47 Proceed by double induction on m and n. If either m or n is zero, both sides are equal to 1 and we have the base. Now let m and n be positive and assume the IH

$$b(m-1, n) = b(n, m-1) \text{ and } b(m, n-1) = b(n-1, m).$$

Then

$$
\begin{aligned}
b(m, n) &= b(m-1, n) + b(m, n-1) \\
&= b(n, m-1) + b(n-1, m) \\
&= b(n-1, m) + b(n, m-1) \\
&= b(n, m).
\end{aligned}
$$

1.49 Double induction again. Choosing m things from m can only be done by taking them all, that is, in just one way. The 0 things that remain can thus be chosen in only one way too. Since $\binom{m+n}{n}$ is equal to 1 when either m or n is zero, this establishes the base. So let m and n be positive and distinguish one object, call it $*$, of the $m+n$ things. Then the m things chosen either include $*$ or they do not. In the first case, include $*$ and choose $m-1$ from the remaining $m+n-1$: $\binom{m+n-1}{m-1}$ choices by the IH. In the second, simply choose m from the remaining $m+n-1$: $\binom{m+n-1}{m}$ choices by the IH again. Thus, the number of ways of choosing m things from $m+n$ is equal to

$$
\begin{aligned}
\binom{m+n-1}{m-1} + \binom{m+n-1}{m} &= b(m-1, n) + (m, n-1) \\
&= b(m, n) \text{ by definition} \\
&= \binom{m+n}{m},
\end{aligned}
$$

as required.

1.50 Divide the $2n$ things into two equal halves. The n things to be chosen will have k in the first half and $n - k$ in the second half, $0 \le k \le n$, giving $\binom{n}{k}$ and $\binom{n}{n-k}$ possibilities respectively. Thus,

$$\binom{2n}{n} = \sum_{k=0}^{n} \binom{n}{k}\binom{n}{n-k} = \sum_{k=0}^{n} \binom{n}{k}^2,$$

by symmetry.

1.53 Adding the equations $P(n)$ and $Q(n)$,

$$
\begin{aligned}
u_{2n} + u_{2n+1} &= u_{n-1}u_n + u_n u_{n+1} + u_n^2 + u_{n+1}^2 \\
&= (u_{n-1} + u_n)u_n + (u_n + u_{n+1})u_{n+1},
\end{aligned}
$$

that is,

$$u_{2n+2} = u_n u_{n+1} + u_{n+1}u_{n+2},$$

which is $P(n + 1)$. Now add $Q(n)$ and $P(n + 1)$:

$$
\begin{aligned}
u_{2n+1} + u_{2n+2} &= u_n^2 + u_{n+1}^2 + u_n u_{n+1} + u_{n+1}u_{n+2} \\
&= u_{n+1}^2 + u_n(u_n + u_{n+1}) + u_{n+1}u_{n+2} \\
&= u_{n+1}^2 + u_n u_{n+2} + u_{n+1}u_{n+2} \\
&= u_{n+1}^2 + (u_n + u_{n+1})u_{n+2},
\end{aligned}
$$

that is,

$$u_{2n+3} = u_{n+1}^2 + u_{n+2}^2,$$

which is $Q(n + 1)$.

The base $n = 1$ of the induction is easy:

$$1 = 0 \cdot 1 + 1 \cdot 1, \qquad 2 = 1^2 + 1^2,$$

and the inductive step has just been proved.

1.54 Induct on n, the formula being obvious when $n = 0, 1$. So let $n \ge 2$. Then

$$
\begin{aligned}
u_n &= u_{n-2} + u_{n-1} \\
&= (\theta^{n-2} - \phi^{n-2})/\sqrt{5} + (\theta^{n-1} - \phi^{n-1})/\sqrt{5} \text{ by the IH} \\
&= (\theta^{n-2} + \theta^{n-1})/\sqrt{5} - (\phi^{n-2} + \phi^{n-1})/\sqrt{5} \\
&= \theta^{n-2}(1 + \theta)/\sqrt{5} - \phi^{n-2}(1 + \phi)/\sqrt{5} \\
&= (\theta^n - \phi^n)/\sqrt{5}
\end{aligned}
$$

as θ and ϕ are the roots of the quadratic $1 + x = x^2$.

1.55 Note that the sum defining c_n for $n \geq 2$ is *symmetric*: the kth term from the left is equal to the kth term from the right. When n is odd, so that the number of terms is even, their sum must be even, whence c_n is even for all odd $n \geq 3$. Further, when $n = 2m$ is even, the parity of c_n is the same as that of the middle term c_m^2, which is the same as the parity of c_m. By putting $n = 2^r.s$ with s odd, and repeating this argument, we see that n is odd precisely when $s = 1$, that is, n is in power of 2.

1.56 Assume the PMI and suppose that the WOP is false, that is, there is a property A that holds for some positive integer, b say, but for no least one. Let $P(n)$ be the proposition:

$$A \text{ fails for every } k \text{ with } 1 \leq k \leq n.$$

Then $P(1)$ is true, for otherwise 1 would be the least positive integer with A. Now let $n > 1$ and assume $P(n-1)$: A fails for every k with $1 \leq k \leq n-1$. Then A must fail for n, otherwise n would be the least positive integer with A. Hence, A fails for all k with $1 \leq k \leq n$. Thus $P(n)$ is true, which completes the inductive step. By the PMI, $P(n)$ is true for all $n \geq 1$, including b. Thus A fails for b. Contradiction.

1.57 We have to show that

$$bs \text{ is an integer } \Leftrightarrow l \mid b.$$

Suppose first that $l \mid b$. Then $b = ml$, and $bs = (ml)s = m(ls)$, and this is an integer since ls is. For the converse, proceed by contraposition and assume that $l \nmid b$. Then $b = ql + r$, $0 < r < l$, by Theorem 1.14. Thus ls is an integer and rs is not. It follows that $bs = q(ls) + rs$ is not an integer, as required.

1.58 Let $a = -c$ with $c > 0$ and write

$$c = bq + r, \qquad 0 \leq r < b.$$

Then

$$a = -c = b(-q) - r = \begin{cases} b(-q) + 0 & \text{if } r = 0, \\ b(-q-1) + (b-r) & \text{if } r > 0, \end{cases}$$

and $0 \leq s < b$ with remainder $s = 0$, $b - r$ in the two cases respectively.

1.60 Put $h = (a, b) = sa + tb$. Since ab/h is a common multiple of a and b, $ab/h \geq [a, b]$. Now let $[a, b] = ma = nb$. Then

$$snb = sma = m(h - tb) = mh - mtb,$$

and $b \mid mh$. Then $b \leq mh$ and $ab \leq amh = [a, b]h$, that is, $ab/h \leq [a, b]$. Hence, $ab = h[a, b] = (a, b)[a, b]$.

1.62
$$\begin{aligned}
a(x) &= (x^3 - x^2)\, b(x) + (x^2 - 1) \\
b(x) &= (x + 1)(x^2 - 1) + (2x + 2) \\
x^2 - 1 &= \tfrac{1}{2}(x - 1)(2x + 2).
\end{aligned}$$

So the hcf is

$$\begin{aligned}
2x + 2 &= b(x) - (x + 1)(x^2 - 1) \\
&= b(x) - (x + 1)(a(x) - (x^3 - x^2)b(x)) \\
&= (x^4 - x^2 + 1)b(x) - (x + 1)a(x).
\end{aligned}$$

(Yes, the algorithm always works for polynomials with rational coefficients, essentially because the *degrees* of the remainders form a strictly decreasing sequence. It is customary to write hcf's of polynomials in *monic* form, that is, with leading coefficient 1, namely, $x + 1$ in this example.)

1.63 Use induction on $n \geq 2$, the inductive base being the content of Theorem 1.16. Now let $n \geq 3$ and assume the result for $n - 1$. Putting $b = b_1 b_2 \cdots b_{n-1}$, we have $p \mid bb_n$, so that $p \mid b$ or $p \mid b_n$ by Theorem 1.16. Thus, by the IH, p divides b_k for some k with $1 \leq k \leq n - 1$ or $p \mid b_n$, as required.

1.64 Since $t_k \leq r_k$ and $t_k \leq s_k$ for all values of k, $1 \leq k \leq l$, it is clear that $h \mid m$ and $h \mid n$. On the other hand, the highest powers of p_k that divide m/h, n/h are $r_k - t_k$, $s_k - t_k$, and one of these is zero by definition of "min". It follows that no prime divides *both* m/h and n/h, whence m/h and n/h are coprime. By Theorem 1.15, we can find integers s and t such that $sm/h + tn/h = 1$. Hence, $sm + tn = h$, and any common divisor of m and n is a divisor of h. Thus $h = (m, n)$.

Chapter 2

2.1 $P \triangle Q \equiv (P \vee Q) \wedge \sim(P \wedge Q)$.

2.2 Yes. Further justification appears in Section 2.2.

2.3 Each side of the first equivalence is false precisely when A and B are both false. Each side of the second equivalence is true precisely when A and B are both true. Hence, both equivalences are true statements.

2.5 C is true if and only if all the P_k are true, $1 \le k \le n$. D is false if and only if all the P_k are false, $1 \le k \le n$.

2.6 For $P \Rightarrow Q$ to be false, we must have P true and Q false. It is true in the other three cases: P and Q both true, P and Q both false, P false and Q true.

2.8 (a) $(\sim P \Rightarrow (Q \wedge \sim Q)) \Rightarrow P$.

   (b) $(\sim Q \Rightarrow \sim P) \Rightarrow (P \Rightarrow Q)$.

   (c) $P(1) \wedge (P(n) \Rightarrow P(n+1)) \Rightarrow P(n)$.

   In (c), the range of values of n in the parentheses on the left and in the $P(n)$ on the right is the whole of $\mathbb{N}$ (see Section 2.4).

2.9 Let $*$ be a unary operation. Then $*P$ can take two truth-values for each truth-value of P, so there are four such operations:

$$
\begin{array}{rcl}
P & : & \text{same truth-value as } P, \text{ the \textbf{identity},} \\
\sim P & : & \text{opposite truth-value to } P, \textbf{ negation,} \\
T = P \vee \sim P & : & \text{always true, \textbf{tautology},} \\
F = P \wedge \sim P & : & \text{always false, \textbf{contradiction}.}
\end{array}
$$

For a binary operation $\circ$, $P \circ Q$ can take two truth-values for each choice of truth-values for P and Q independently. Since there are $2 . 2 = 4$ such choices, the number of possibilities for $\circ$ is $2^4 = 16$.

Pursuing similar reasoning, the total number of n-any operations is 2^{2^n}.

2.11 The left-hand side is the property, call it P, of n in Theorem 1.16, and the right-hand side the definition of prime or 1. Thus, "if n has property P, then n is prime or $n = 1$". This is the converse of Theorem 1.16.

2.12 It is a form of variant 3 of the PMI (double induction). As in Exercise 2.8, the ranges of values of parameters are missing (see Section 2.4).

2.13

P	$\sim P$	T	F
1	0	1	0
0	1	1	0

2.14

P	$\equiv$	P
1	1	1
0	1	0

P	$\vee$	$\sim$	P
1	1	0	1
0	1	1	0

$\sim$	$(P$	$\wedge$	$\sim$	$P)$
1	1	0	0	1
1	0	0	1	0

2.17

$(P$	$\Rightarrow$	$(Q$	$\Rightarrow$	$R))$	$\equiv$	$(Q$	$\Rightarrow$	$(P$	$\Rightarrow$	$R))$
1	1	1	1	1	1	1	1	1	1	1
1	0	1	0	0	1	1	0	1	0	0
1	1	0	1	1	1	0	1	1	1	1
1	1	0	1	0	1	0	1	1	0	0
0	1	1	1	1	1	1	1	0	1	1
0	1	1	0	0	1	1	1	0	1	0
0	1	0	1	1	1	0	1	0	1	1
0	1	0	1	0	1	0	1	0	1	0

2.19

$(P$	$\Rightarrow$	$Q)$	$\Rightarrow$	$((Q$	$\Rightarrow$	$R)$	$\Rightarrow$	$(P$	$\Rightarrow$	$R))$
1	1	1	1	1	1	1	1	1	1	1
1	1	1	1	1	0	0	1	1	0	0
1	0	0	1	0	1	1	1	1	1	1
1	0	0	1	0	1	0	0	1	0	0
0	1	1	1	1	1	1	1	0	1	1
0	1	1	1	1	0	0	1	0	1	0
0	1	0	1	0	1	1	1	0	1	1
0	1	0	1	0	1	0	1	0	1	0

2.22   The tautology $T = P \vee \sim P$ is defined by four ones. Each of $P \vee Q$, $P \vee \sim Q$, $\sim P \vee Q$ has exactly three ones. Each of $P, Q, (P \wedge Q) \vee (\sim P \wedge \sim Q)$ has exactly two ones. $P \wedge Q$ has just one one.

Since these eight operations are all different, they comprise half of the 16. Since they are all true when P and Q are both true, their negations comprise the other half.

2.24 CELARENT first figure: no b is a c every a is a b $\therefore$ no a is a c.

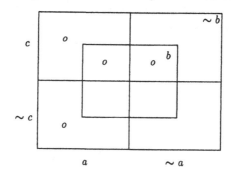

2.27 Some positive integers are not even: true

every positive multiple of 4 is a positive integer: true

some positive multiples of 4 are not even: false.

2.30 Take BARBARI in the first figure, for example:

every b is a c

every a is a b

some a is a c

Certainly, every a is a c (BARBARA), so how could "some a is a c" fail? Since "some" means "at least one", only when a is *empty*, for example, "integer root of the equation $x^2 + x + 1 = 0$". The modern convention is always to allow emptiness (see Chapter 3), just as 0 is allowed to be an integer.

2.31

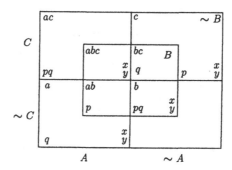

We first enter the lower case letter for each of A, B, C in just those sections where the corresponding upper-case letter is true. Then we put a p in those sections containing exactly one of b and c, and similarly with q for a and b. Finally, the same again with x, y for a and p, q and c, respectively. Since the sections containing x are the same as those containing y, we deduce that $X \equiv Y$, that is, $A_\triangle(B_\triangle C) \equiv (A_\triangle B)_\triangle C$.

2.32

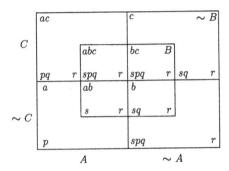

First enter a's, b's, c's as in the previous exercise. Take S to be $A \Rightarrow B$, that is, $\sim A \vee B$, and enter s in every section containing b or not containing a. Similarly for $P = \sim B \vee C$ and $Q = \sim A \vee C$. Finally, do the same for $R = P \Rightarrow Q$. Since every section containing s also contains r, we deduce that $S \Rightarrow R$ as required.

2.34 (a) The implication is true: just take $x = 1$.

(b) For this to be true, there must be at least one x in the range of its values. So it is true in all cases except when that range is empty (cf. Exercise 2.30 and its solution above).

2.36 Assume the left-hand side, and let a be a value of x whose existence is asserted. Then $P(a, y)$ is true $\forall y$, and we can take $x = a \ \forall y$ on the right-hand side.

2.38 Let $R(n)$ be the proposition that formula (9) is true ($\forall P(\cdots)$ and $\forall$ combinations of Q_k, $1 \leq k \leq n$) and induct on n. When $n = 1$ the proposition is the assertion of Theorem 1, which we know to be true. Now let $n \geq 1$ and assume $R(n)$ as the IH. Then abbreviating $P(x_1, \ldots, x_{n+1})$ to P for convenience,

$$\sim \prod_{k=1}^{n+1} Q_k x_k P \equiv \sim \prod_{k=1}^{n} Q_k x_k Q_{n+1} x_{n+1} P$$

$$\equiv \prod_{k=1}^{n} \overline{Q}_k x_k \sim Q_{n+1} x_{n+1} P \qquad \text{by IH}$$

$$\equiv \prod_{k=1}^{n} \overline{Q}_k x_k \, \overline{Q}_{n+1} x_{n+1} \sim P \qquad \text{by } R(1)$$

$$\equiv \prod_{k=1}^{n+1} \overline{Q}_k x_k \sim P$$

which proves $R(n+1)$. This completes the inductive step

2.39 Any two positive integers have a least common multiple.

2.40 $\exists a\, \exists b\, \forall c\, (a \nmid c) \vee (b \nmid c) \vee (a \mid d \wedge b \mid d \wedge c > d)$.

2.42 Positive rational numbers.

Chapter 3

3.1 $P(1,1) \wedge (\forall m, n \in \mathbb{N}\ P(m,n) \Rightarrow (P(m+1,n) \wedge P(m,n+1))) \Rightarrow \forall m, n \in \mathbb{N}\ P(m,n)$,

$\forall n \in \mathbb{N}\ \exists m \in \mathbb{N}\ m > n$,

$\exists m \in \mathbb{N}\ \forall n \in \mathbb{N}\ m > n$.

(Recall that the first two are true, but the third is false.)

3.2 (i) The vertical axis, or y-axis,

  (iv) the (lattice of) integer points.

3.3 (ii) a parabola if $a \neq 0$, a (straight) line otherwise,

  (iv) a square.

3.5 We have $x \in A' \Leftrightarrow x \notin A$. Negating both sides, $x \notin A' \Leftrightarrow x \in A$. Replacing A by A', $x \notin (A')' \Leftrightarrow x \in A' \Leftrightarrow x \notin A$. Negating the left- and right-hand ends of this,

$$x \in (A')' \Leftrightarrow x \in A.$$

So $(A')' = A$.

3.7 We have $x \in A \Rightarrow x \in B \Rightarrow x \in C$. So $x \in A \Rightarrow x \in C$, that is, $A \subseteq C$.

3.8 $C \subseteq B' \wedge B \not\subseteq A' \Rightarrow A \not\subseteq C$.

By the minor premise, there is an element $a \in A$ with $a \in B$. Then $a \notin C$, by the major premise. So $a \in A$ and $a \notin C$ so that $A \not\subseteq C$, as required.

3.9 CALIBAN in the second figure asserts that

$$C \subseteq B \wedge A \not\subseteq B' \Rightarrow A \subseteq C?$$

To invalidate it, we need an example of subsets A, B, C of some set S with

$$C \subseteq B \wedge A \not\subseteq B' \wedge A \not\subseteq C.$$

Take $S = A = B = \{1\}$, so $B' = \emptyset$, and $C = \emptyset$

3.10 $S_{m,n} \subseteq S_{p,q} \Leftrightarrow (m \geq p) \wedge (n \leq q)$.

3.12 Oscillate outwards from the middle: if $z \in \mathbb{Z}$ is positive, give it the number $2z$ and if $z \leq 0$, give it the number $1 - 2z$.

3.13 First number quotients m/n for $m, n \in \mathbb{N}$ by increasing values of $m + n = 2, 3, 4, \ldots$, starting in each case with $m = 1$. So first eight quotients are $1/1, 1/2, 2/1, 1/3, 2/2, 3/1, 1/4, 2/3$. Then throw away all quotients m/n with $(m,n) \neq 1$.

3.15 $\cap, \cup, \triangle$ are commutative. If $S = \{1,2\}$, $A = \{1\}$ and $B = \{2\}$, then

$$A \setminus B = \{1\} = A \neq B = \{2\} = B \setminus A.$$

3.16 $\forall A \subseteq S$, $A \cup \emptyset = A$, so $\emptyset$ is an identity for $\cup$.

$\forall A \subseteq S$, $A \cap S = A$, so S is an identity for $\cap$.

Let $S \neq \emptyset$ and assume that B is an inverse for S under $\cup$, that is, $S \cup B = \emptyset$. But $S \subseteq S \cup B$, so $S = \emptyset$, contradiction.

Again let $S \neq \emptyset$ and assume that B is an inverse for $\emptyset$ under $\cap$, that is, $\emptyset \cap B = S$. But $\emptyset \cap B = \emptyset$, so $S = \emptyset$, a contradiction.

3.18 $\begin{aligned}[t] x \in (A \cup B)' \quad &\Leftrightarrow \quad \sim(x \in A \cup B) \\ &\Leftrightarrow \quad \sim(x \in A \lor x \in B) \\ &\Leftrightarrow \quad (\sim x \in A) \land (\sim x \in B), \quad \text{by Exercise 2.15,} \\ &\Leftrightarrow \quad (x \in A') \land (x \in B') \\ &\Leftrightarrow \quad x \in A' \cap B'. \end{aligned}$

So $(A \cup B)' = A' \cap B'$. For the other law:

replace A, B by A', B' respectively in this formula, apply $'$ to both sides, and use the fact that $(C')' = C$ three times.

3.20 Let $A = \{a_i \mid 1 \leq i \leq m\}$ and $B = \{b_j \mid 1 \leq j \leq n\}$. Then $A \cup B$ consists of the distinct elements in the list

$$a_1, a_2, \ldots, a_n, b_1, b_2, \ldots, b_n.$$

But no two of these are the same as $A \cap B = \emptyset$. So $|A \cup B| = m + n = |A| + |B|$.

3.21 Since $A \setminus B$ and B are disjoint,

$$|A \cup B| = |A \setminus B| + |B|.$$

Since $A \cap B$ and $A \setminus B$ are disjoint,

$$|A| = |A \cap B| + |A \setminus B|.$$

Eliminating $|A \setminus B|$ from these two equations gives the result.

3.22 Put

$$\begin{aligned} A = A_1 \setminus A_2, \quad &\text{so} \quad |A| = 2, \\ B = A_3 \setminus A_2, \quad &\text{so} \quad |B| = 11. \end{aligned}$$

Then, by the previous exercise, $|A \cup B| = 11, 12$, or 13. But $|A_1 \cup A_2 \cup A_3| = |A \cup B| + |A_2|$ since $A \cup B$ and A_2 are disjoint. Since $|A_2| = 15$, we get the result.

3.23 Set $A = \{a_i \mid 1 \le i \le m\}, B = \{b_j \mid 1 \le j \le n\}$. Since the ordered pairs (a_i, b_j), $1 \le i \le m$, $1 \le j \le n$, are all distinct, they comprise the set $A \times B$. Then $|A \times B|$ is equal to the number of these pairs, that is, mn.

3.26 $A, A', S, \emptyset$.

3.27

(A	\	B)	∪	(B	\	A)	=	(A	∪	B)	\	(A	∩	B)
1	0	1	0	1	0	1	1	1	1	1	0	1	1	1
1	1	0	1	0	0	1	1	1	1	0	1	1	0	0
0	0	1	1	1	1	0	1	0	1	1	1	0	0	1
0	0	0	0	0	0	0	1	0	0	0	0	0	0	0

3.29
$$|A_1 \cup A_2 \cup A_3| = |A_1| + |A_2 \cup A_3| - |A_1 \cap (A_2 \cup A_3)|$$
$$= |A_1| + |A_2| + |A_3| - |A_2 \cap A_3|$$
$$- |(A_1 \cap A_2) \cup (A_1 \cap A_3)|$$
$$= |A_1| + |A_2| + |A_3| - |A_2 \cap A_3|$$
$$- (|A_1 \cap A_2| - |A_1 \cap A_3| + |A_1 \cap A_2 \cap A_1 \cap A_3|).$$

This is the same as the right-hand side, since

$$A_1 \cap A_2 \cap A_1 \cap A_3 = A_1 \cap A_2 \cap A_3,$$

by associativity and commutativity of $\cap$ together with the (obvious) **idempotent law** $A \cap A = A$.

3.30

A	=	B
1	1	1
1	0	0
0	0	1
0	1	0

A	⊆	B
1	1	1
1	0	0
0	1	1
0	1	0

3.35 Five, with members

$S,$

$\{1,2\}, \{3\},$

$\{1,3\}, \{2\},$

$\{2,3\}, \{1\},$

$\{1\}, \{2\}, \{3\},$

respectively

3.37 Take $T = \mathbb{Z}$.

3.38 You should get 3, 5, 7, 11 respectively. The guess $P_7 = 13$ is wrong; in fact, $p_7 = 15$.

3.39 Just four, with members

$\emptyset, S,$

$\emptyset, \{1\}, S,$

$\emptyset, \{2\}, S,$

$\emptyset, \{1\}, \{2\}, S,$

respectively.

3.42 Check conditions (i)–(iv) in turn.

   (i) Since every topology τ has $\emptyset$ as a member, $|\tau| \geqslant 1$. Hence, $\emptyset$ is not a topology as $|\emptyset| = 0$. So $\emptyset \notin T$, (i) fails, and T is not a topology. We continue for the sake of completemers (and curiosity).

   (ii) $\tau = \mathcal{P}(S)$ is a topology (obviously), called the **discrete topology** on S, so (ii) holds.

   (iii) Consider $\sigma \cup \tau$, where σ and τ are topologies on S. (i) and (ii) hold, but (iii) and (iv) fail in general: take $S = \{1, 2, 3\}$, $\sigma = \{\phi, \{1\}, \{1, 2\}, S\}$, $\tau = \{\phi, \{3\}, \{3, 2\}, S\}$ and observe that neither $\{1\} \cup \{3\} = \{1, 3\}$ nor $\{1, 2\} \cap \{3, 2\} = \{2\}$ is in $\sigma \cup \tau$. So (iii) fails.

   (iv) Let $\sigma, \tau \in T$ and consider $\sigma \cap \tau$. It is easily checked that (i)–(iv) hold for $\sigma \cap \tau$, so that (iv) holds for T.

Chapter 4

4.1 (S) holds, but (R) and (T) fail.

4.3 (R) is obvious, and (S) follows from the symmetry of $=$.

   For (T), assume $(a, b) \sim (c, d)$ and $(c, d) \sim (e, f) : a + d = c + b$, $c + f = e + d$.

   Adding these together and subtracting $c + d$ from both sides, we get $a + f = e + b$, making free use of the fundamental laws of arithmetic. So $(a, b) \sim (e, f)$, as required.

4.5 For (R), take $h(x) = 0$.

   For (S), replace $h(x)$ by $-h(x)$.

   For (T), add the multiples together.

4.6 The statements are not quantified properly: this "proof" starts by apparently assuming that $a \rho b$ for some b, and it may be that there is no such b, for example when ρ is the "empty relation" $\emptyset$.

4.7 This is the relation $=$ of equality.

4.9 (R), (S), (T) follow from the corresponding properties of $=$. The classes are the half-open intervals $[n, n + 1), n \in \mathbb{Z}$, in formula (3.8) of the previous chapter. $\mathbb{Z}$ is a transversal.

4.10 By the one-to-one correspondence, we can take the bound of Exercise 3.36 for the number of partitions, namely 2^{2^n}.

4.13 In each of these five laws, the right- and left-hand sides are equal for all integer values of the variables. Hence, in each case, their remainders on division by n are equal.

4.15 Suppose that n is not a prime, $n = a.b$ say, where $a, b \in \mathbb{N}$, $1 < a, b < n$. Suppose further, for a contradiction, that $\mathbb{Z}/n\mathbb{Z}$ is a field. So $\forall [a] \neq [0]$, $[a]$ has a multiplicative inverse, call it $[c]$, in $\mathbb{Z}/n\mathbb{Z}$. Then

$$
\begin{aligned}
[ca] = [c][a] = [1] \quad &\Rightarrow \quad ca \equiv 1 \pmod{n} \\
&\Rightarrow \quad n \mid (ca - 1) \\
&\Rightarrow \quad n \mid (ca - 1)b = cn - b \\
&\Rightarrow \quad n \mid b,
\end{aligned}
$$

   which is impossible as $1 < b < n$. So $\mathbb{Z}/n\mathbb{Z}$ is not a field.

4.16 Recall from Chapter 1 that a field is just a commutative ring-with-1 in which every non-zero element has a multiplicative inverse. Suppose that n is a prime. Then we must prove that

$$[0] \neq [a] \in \mathbb{Z}/n\mathbb{Z} \Rightarrow \exists c \, [ca] = [c][a] = [1].$$

Since $[0] \neq [a]$, we can assume that $1 \leq a \leq n - 1$. Then a and n are coprime, $(a, n) = 1$, as n is a prime. By Theorem 1.15, $\exists s, t \in \mathbb{Z}$ $1 = sa + tn$. Then $[sa] = [1]$ and we can take $c = s$.

4.17 If $(a, n) = 1$, the multiplicative inverse $[c]$ is found exactly as in the previous solution. For the converse, suppose that $[a, n] = h \neq 1$ so that h divides every multiple of a. Assume for a contradiction that $[a]$ has a multiplicative inverse $[c] : [ca] = [c][a] = 1$. Then $ca - 1 = kn$ for some $k \in \mathbb{Z}$. Since $h \mid a$ and $h \mid n$, we deduce $h \mid 1$, which is impossible as $h > 1$. So no multiplicative inverse exists when $(a, n) \neq 1$.

4.20 Suppose that $(a', b') \sim (a, b)$, $(c', d') \sim (c, d)$, that is, $a' + b = a + b'$, $c' + d = c + d'$. Adding these equations, $(a' + c') + (b + d) = (a + c) + (b' + d')$, that is, $(a' + c', b' + d') \sim (a + c, b + d)$, as required.

4.21 Multiplying $-1 = [1, 2]$ by $-b = [1, b + 1]$ using (4.11),

$$\begin{aligned}
[1, 2][1, b + 1] &= [1 . 1 + 2(b + 1), 1 . (b + 1) + 2 . 1] \\
&= [2b + 3, b + 3] \\
&= [b + 1, 1],
\end{aligned}$$

subtracting $b + 2$ from each component, and $[b + 1, 1] = b$.

4.23 Using (4.11),

$$\begin{aligned}
[2, 1][a, b] &= [2a + b, 2b + a] \\
&= [a, b],
\end{aligned}$$

subtracting $a + b$ from both components.

4.24 We have to show that none of the products

$$[m+1, 1][n+1, 1], [m+1, 1][1, n+1], [1, m+1][n+1, 1], [1, m+1][1, n+1]$$

is equal to the zero class $[1, 1] = \{(k, k) \mid k \in \mathbb{N}\}$. Using (4.11) and subtracting $m + n + 1$ from each component, these products are $[mn + 1, 1], [1, mn + 1], [1, mn + 1], [mn + 1, 1]$ respectively. Since $mn + 1 \neq 1$, none of them is equal to $[1, 1]$.

4.25 Suppose that $(a', b') \sim (a, b)$, $(c', d') \sim (c, d)$, that is,

$$a'b = ab', \quad c'd = cd'.$$

Multiplying these equations,

$$a'c'bd = acb'd',$$

that is, $(a'c', b'd') \sim (ac, bd)$, as required.

4.26 $\begin{aligned}[a, b] + [-a, b] &= [ab - ba, b^2] \\ &= [0, b^2] \\ &= [0, 1].\end{aligned}$

4.30 Given $k \in \mathbb{N}$, let $N \in \mathbb{N}$ be such that $k < 10^N$. Then $\forall m, n \in \mathbb{N}$ with $m \geq n > N$,

$$-1/k < 0 \leq a_m - a_n = \sum_{i=n+1}^{m} d_i 10^{-i} = 10^{-n} \sum_{i=n+1}^{m} d_i 10^{n-i}$$

$$\leq 10^{-n} \sum_{i=n+1}^{m} 9 \times 10^{n-i} < 10^{-n}$$

$$< 10^{-N} < 1/k,$$

as required by condition (4.19).

d_k is the kth digit after the decimal point in the decimal expansion of r.

4.31 If $a = b$, then $a \rho b$ by (R). Conversely, if $a \rho b$, then $b \rho a$ by (S). But then $a = b$ by (O). So ρ can only be equality.

4.32 (R): if $a = b$ then $a \rho b$ by definition.

$$\begin{aligned}(\mathrm{O}) : a \rho b \wedge b \rho a &\Rightarrow (a \sigma b \vee a = b) \wedge (b \sigma a \vee b = a) \\ &\Rightarrow (a \sigma b \wedge b \sigma a) \vee (a = b) \\ &\Rightarrow (a \sigma a) \vee (a = b) \\ &\Rightarrow a = b,\end{aligned}$$

using in turn the definition, a distributive law, (T) for σ, (I) for σ. Similarly,

$$\begin{aligned}(\mathrm{T}) : a \rho b \wedge b \rho c &\Rightarrow (a \sigma b \vee a = b) \wedge (b \sigma c \vee b = c) \\ &\Rightarrow (a \sigma b \wedge b \sigma c) \vee (a \sigma b \wedge b = c) \\ &\quad \vee (a = b \wedge b \sigma c) \vee (a = b \wedge b = c) \\ &\Rightarrow (a \sigma c) \vee (a \sigma c) \vee (a \sigma c) \vee (a = c) \\ &\Rightarrow a \rho c.\end{aligned}$$

4.33 Assume that ρ satisfies (W). To prove (L), let a and b be elements of S. First, if $a = b$, (R) asserts that $a \rho b$ holds. If $a \neq b$, apply (W) to the non-empty set $A = \{a, b\}$. Then, for the least element l, $l = a \vee l = b$, yielding $a \rho b \vee b \rho a$, as required.

4.35 By definition, every $P \in \Pi$ is non-empty. By (W), each P has a least element, call it l_P. Then $\{l_P \mid P \in \Pi\}$ is a transversal.

4.36 For all $m, n \in \mathbb{N}$,

$$-m \leq 0 \leq n, \ m \leq n \Rightarrow -n \leq -m.$$

For $a/b, c/d \in \mathbb{Q}$ with $bd \neq 0$, define

$$a/b \leq c/d \Leftrightarrow ad - bc \leq 0.$$

4.38 Put $r = [a_n]$, where (a_n) is a Cauchy sequence of rational numbers. Define

$$r \geq 0 \quad \Leftrightarrow \quad \exists N \in \mathbb{N} \ \forall n > N \ a_n \geq 0.$$

Chapter 5

5.1 (a) For negative values of the argument there is no real square-root, and

    (b) for positive values there are two.

By definition, every element of the domain must pass to exactly one value under a map.

5.2 There may be distinct elements $a_1, a_2 \in A$ such that $\theta(a_1) = \theta(a_2)$, whereupon $\{\theta(a) \mid a \in A\}$ is not a set in accordance with the definition.

5.3 Their graphs consist of the following pairs of elements of $A \times B$:

$$(a,1),(b,1); \qquad (a,1),(b,2); \qquad (a,1),(b,3);$$
$$(a,2),(b,1); \qquad (a,2),(b,2); \qquad (a,2),(b,3);$$
$$(a,3),(b,1); \qquad (a,3),(b,2); \qquad (a,3),(b,3).$$

5.5 $\gamma : \mathrm{Map}(A, B) \longrightarrow \mathcal{P}(A \times B)$.

$\mathrm{Im}\,\gamma = \{S \subseteq A \times B \mid \forall a \in A\ \exists\mid b \in B\ (a,b) \in S\}$.

5.6 Since $A \neq \emptyset$, $\exists a \in A$. Since there is no value we can assign to a, $\mathrm{Map}(A, \emptyset)$ is the empty set.

Since $\emptyset \subseteq B$ for any set B, there is an inclusion $\iota \in \mathrm{Map}(\emptyset, B)$, which is thus a singleton. The same argument holds when $B = \emptyset$, so $\mathrm{Map}(\emptyset, \emptyset)$ is a singleton.

5.7 $K = \{(a,c) \in A \times C \mid \exists b \in B\ \ (a,b) \in G \wedge (b,c) \in H\}$.

5.9 $\forall X \subseteq A$, put $\theta(X) = \{b \in B \mid \exists x \in X\ \theta(x) = b\}$. Since $A \neq \mathcal{P}(A)$, the same symbol θ is being used for two different maps. This is all right *provided* any use of the symbol θ (as in Exercise 5.10) is accompanied by a qualification explaining which of the two meanings is intended.

5.11 Since $A \neq \emptyset$ there is at least one element, call it a, in A. Then define

$$
\begin{aligned}
\sigma : \quad B &\longrightarrow A \\
b &\longrightarrow \begin{cases} b & \text{if } b \in A, \\ a & \text{if } b \notin A. \end{cases}
\end{aligned}
$$

5.13 Any $q \in Q$ is equal to a $\sim$-class, call it $[a]$, in A. We would like to define $\theta'([a]) = [\theta(a)]'$, the $\sim'$-class of $\theta(a)$ in Q', but must check for independence of representatives (cf. Chapter 4, Section 3 above). Since

$$b \in [a] \Rightarrow a \sim b \Rightarrow \theta(a) \sim' \theta(b) \Rightarrow [\theta(a)]' = [\theta(b)]',$$

it follows that θ' is well defined. Then $\forall a \in A$,

$$\theta'\nu(a) = \theta'[a] = [\theta(a)]' = \nu'\theta(a).$$

5.14 Another hairy one. Denoting the $\sim$-class of a by $[a]$, we might try $[a]*[b] = [a*b]$, $a, b \in A$. This is well defined as long as for $a, a', b, b' \in A$,

$$a \sim a', \; b \sim b' \Rightarrow a*b \sim a'*b'.$$

5.16 As $f^{-1}(0)$ is the set of roots of the equation $f(x) = 0$, $|f^{-1}(0)| \leq n$.

5.17 If $\beta : B \times B \longrightarrow B$ is the binary operation and $\delta : A \longrightarrow A \times A$ is the diagonal map, then $\theta * \phi$ is the composite $\beta(\theta \times \phi)\delta$.

5.18 Given $\theta, \phi, \psi \in \text{Map}(A, A)$, we may have

$$\theta(\phi * \psi) = (\theta\phi) * (\theta\psi), \qquad (\phi * \psi)\theta = (\phi\theta) * (\psi\theta),$$
$$\theta * (\phi\psi) = (\theta * \phi)(\theta * \psi), \qquad (\phi\psi) * \theta = (\phi * \theta)(\psi * \theta).$$

Only the second one is always true: $\forall a \in A$

$$(\phi * \psi)\theta(a) = (\phi\theta)(a) * (\psi\theta)(a) = (\phi\theta * \psi\theta)(a).$$

5.19 **A sequence** of real numbers.

5.21 Just six of the nine, those with graphs:

$$(a, 1), \; (b, 2); \; (a, 1), \; (b, 3); \; (a, 2), \; (b, 1); \; (a, 2), \; (b, 3); \; (a, 3),$$
$$(b, 1); \; (a, 3), \; (b, 2).$$

None are surjections.

5.23 For $x, y \in \mathbb{R}$,

$$\theta(x) = \theta(y) \Leftrightarrow ax + b = ay + b \Leftrightarrow ax = ay \Leftrightarrow a(x - y) = 0.$$

We can deduce $x = y$ from this if and only if $a \neq 0$.

For any $c \in \mathbb{R}$, $\exists x \in \mathbb{R}$ with

$$\theta(x) = c \Leftrightarrow ax + b = c \Leftrightarrow ax = c - b,$$

and this can be solved for x $\forall c$ if and only if $a \neq 0$. The condition for (i), (ii) or (iii) is thus the same in every case: $a \neq 0$.

5.24 The values of ϕ on 1, 2, 3, 4 are 0, 1, 2, 2, respectively, so ϕ is not injective: $\phi(3) = \phi(4)$. Since $\forall n \in \mathbb{N}$ $\phi(n) \geq 0$, ϕ is not surjective: $\nexists n \in \mathbb{N}$ $\phi(n) = -1$. (It turns out that $\text{Im}\,\phi \subseteq \{0, 1\} \cup \{2n \mid n \in \mathbb{N}\}$.)

5.27 If $m > n$, there are no injections by condition (5.20). So assume $m \leq n$ and put $A = \{a_k \mid 1 \leq k \leq m\}$. The values of an injection on these elements can be chosen in turn: for a_1 in n ways, for a_2 in $n - 1$ ways, and when $k \geq 3$, for a_k in $n-k+1$ ways, finishing with $n-m+1$ choices when $k = m$. The number of injections is thus equal to $n!/(n - m)!$

The number of bijections is equal to $n!$ when $m = n$ and 0 otherwise.

It is quite difficult (but not impossible) to count the number of surjections.

5.28 Let $b \in \operatorname{Im} \theta$, so that $\theta^{-1}(b) \neq \emptyset$ and $\theta^{-1}(b) \in Q$. Then define

$$\theta' : \quad \begin{array}{ccc} Q & \longrightarrow & \operatorname{Im} \theta \\ \theta^{-1}(b) & \longmapsto & b \end{array} \quad ,$$

which is clearly well defined. Since θ' has an inverse:

$$\begin{array}{ccc} \operatorname{Im} \theta & \longrightarrow & Q \\ b & \longmapsto & \theta^{-1}(b) \end{array} \quad ,$$

it is a bijection.

5.29 With θ, A, B, Q, θ' as in the previous exercise, we have a natural map $\nu : A \longrightarrow Q$ and an inclusion $\iota : \operatorname{Im} \theta \longrightarrow B$. Then $\theta = \iota \theta' \nu$ is the required decomposition.

5.31 Since $(A')' = A \; \forall A \subseteq S$, it follows that $\kappa \kappa = 1_{\mathcal{P}(S)}$. So κ has an inverse (κ itself), and is thus a bijection.

5.33 The map

$$\begin{array}{ccc} \mathcal{P}(A \dot\cup B) & \longrightarrow & \mathcal{P}(A) \times \mathcal{P}(B) \\ C & \longmapsto & (C \cap A, C \cap B) \end{array}$$

has an inverse

$$\begin{array}{ccc} \mathcal{P}(A) \times \mathcal{P}(B) & \longrightarrow & \mathcal{P}(A \dot\cup B) \\ (C, D) & \longmapsto & C \cup D \end{array}$$

5.35 Given any map $\theta : A \times B \longrightarrow C$ and $a \in A$, define $\theta_a : B \longrightarrow C$ by setting $\theta_a(b) = \theta(a, b)$. Then define the image of θ under β to be the map$(a \longmapsto \theta_a)$ from A to $\operatorname{Map}(B, C)$. Next, given a map $\phi : A \longrightarrow \operatorname{Map}(B, C)$, let $\gamma(\phi)$ be the map from $A \times B \longrightarrow C$ sending (a, b) to $(\phi(a))(b)$ ($\phi(a)$ is a map from B to C). Then a simple check shows that β and γ are mutually inverse, so β is a bijection.

5.36 Suppose for a contradiction that N is finite. Then by the pigeon-hole princliple, $P1 \Rightarrow \sigma$ is onto $\Rightarrow 1 \in \operatorname{Im} \sigma$, which contradicts P2. Therefore N is an infinite set.

5.38 Put $A = \{a \in N \mid a + 1 = 1 + a\}$ and apply P3. Clearly $1 \in A$. Let $a \in A$. Then

$$
\begin{aligned}
a^* + 1 &= (a+1) + 1 & \text{By A1} \\
&= (1+a) + 1 & \text{as } a \in A \\
&= 1 + (a+1) & \text{by Exercise 5.37} \\
&= 1 + a^*.
\end{aligned}
$$

Hence, $a^* \in A$ and P3 $\Rightarrow A = N$, as required.

5.40 Put $X = \{c \in N \mid a \longmapsto a + c \text{ is an injection}\}$. Then $1 \in X$ by P1 and A1. Let $c \in X$ and consider the map: $N \longrightarrow N, a \longmapsto a + c^*$. Then

$$
\begin{aligned}
a + c^* = a' + c^* &\Rightarrow (a+c)^* = (a'+c)^* & \text{by A2} \\
&\Rightarrow a + c = a' + c & \text{by P1} \\
&\Rightarrow a = a' & \text{since } c \in X.
\end{aligned}
$$

Thus, $a \longmapsto a + c^*$ is also an injection, that is, $c^* \in X$. Hence, $X = N$ and the cancellation law holds.

5.45 Let $A = \{a \in N \mid \forall b \in N \; ba + a = (b+1)a\}$. Clearly $1 \in A$, by M1 (twice). Let $a \in A$. Then

$$
\begin{aligned}
ba^* + a^* &= (ba + b) + (a+1) & \text{by M2 and A2} \\
&= (ba + a) + (b+1) & \text{by Theorem 5.9(a), (b)} \\
&= (b+1)a + (b+1) & \text{since } a \in A \\
&= (b+1)a^* & \text{by M2.}
\end{aligned}
$$

So $a^* \in A$ and $A = N$.

5.49

$$
\begin{aligned}
a < b &\Rightarrow \exists x \in N \; a + x = b, \\
&\Rightarrow \exists x \in N \; \forall c \in N \; c + a + x = c + b, \\
&\Rightarrow \forall c \in N \; c + a < c + b, & \text{the second assertion,} \\
&\Rightarrow 1 + a < 1 + b, \\
&\Rightarrow \exists c \in N \, c + a < c + b, & \text{the third assertion,} \\
&\Rightarrow \exists c, x \in N \, c + a + x = c + b, \\
&\Rightarrow \exists x \in N \, a + x = b, & \text{by Theorem 5.9(c),} \\
&\Rightarrow a < b.
\end{aligned}
$$

Chapter 6

6.1 $\theta(a) = (a, 0)$, $\phi(b) = (b, 1)$.

6.2 Given $\mathbf{a} = (a_1, a_2, \ldots, a_n) \in A^n$, define $\alpha : S \longrightarrow A$ by $\alpha(i) = a_i$, $1 \le i \le n$, and write $\theta(\mathbf{a}) = \alpha$. Given $\alpha : S \longrightarrow A$, define $\phi(\alpha) = (\alpha(1), \alpha(2), \ldots, \alpha(n)) \in A^n$. Then θ and ϕ are mutually inverse, so θ is a bijection.

6.3 Déjà vu!

6.4 Given bijections $\alpha : A \longrightarrow \overline{A}$, $\beta : B \longrightarrow \overline{B}$, define

$$
\begin{aligned}
\gamma : A \dot{\cup} B \quad &\longrightarrow \quad \overline{A} \dot{\cup} \overline{B} \\
x \quad &\longmapsto \quad \begin{cases} \alpha(x) & \text{if } x \in A, \\ \beta(x) & \text{if } x \in B. \end{cases}
\end{aligned}
$$

This map has an obvious inverse, and so is a bijection.

6.5 Given bijections $\alpha : A \longrightarrow \overline{A}$, $\beta : B \longrightarrow \overline{B}$, let $\gamma = \alpha \times \beta$. Then $\gamma^{-1} = \alpha^{-1} \times \beta^{-1}$.

6.7 These laws follow at once from the corresponding ones for sets with respect to $\dot{\cup}$ and $\times$.

6.9 (a) The Axiom of Choice, to get a transversal.

(b) $B \ne \emptyset$, to get the map.

6.10 Let A, B be sets with $|A| = a$, $|B| = b$. Then we seek an injection $\gamma : \mathrm{Map}(A, B) \longrightarrow \mathcal{P}(A \times B)$. Given $\theta : A \longrightarrow B$, define $\gamma(\theta)$ to be the graph of θ,

$$
\gamma(\theta) = \{(a, b) \in A \times B \mid \theta(a) = b\}.
$$

6.12 The maps

$$
\begin{array}{ccc}
(0, 1) \longrightarrow (a, b) & \qquad & (a, b) \longrightarrow (0, 1) \\
x \longmapsto a + x(b - a) & , & y \longmapsto (y - a)/(b - a)
\end{array}
$$

are mutually inverse, whence $(0, 1) \sim (a, b)$ $\forall a, b \in \mathbb{R}$ with $a < b$. The result now follows by (S) and (T).

6.13 $X \subseteq Y \implies \theta(X) \subseteq \theta(Y) \implies \theta(X)' \supseteq \theta(Y)'$
$\qquad \implies \phi(\theta(X)') \supseteq \phi(\theta(Y)')$
$\qquad \implies \omega(X) = \phi(\theta(X)')' \subseteq \phi(\theta(Y)')' = \omega(Y)$.

6.14 Since $\emptyset \subseteq$ any set, $\emptyset \subseteq \omega(\emptyset)$, whence $\emptyset \in \mathcal{F}$. On the other hand,

$$
\begin{aligned}
A \in \mathcal{F} \;&\Longrightarrow\; A \subseteq \omega(A) \Longrightarrow A = \omega(A) \\
&\Longrightarrow\; \mathrm{Im}\,\phi\alpha\theta = \emptyset \Longrightarrow \mathrm{Im}\,\alpha\theta = \emptyset \\
&\Longrightarrow\; \mathrm{Im}\,\theta = B \Longrightarrow \theta \;\text{ is surjective.}
\end{aligned}
$$

6.15 For any $X \subseteq \mathbb{N}$, define $X + 1 = \theta(X) = \{x + 1 \mid x \in X\}$.

Then $\mathbb{N} = X \dot{\cup} X' \;\Rightarrow\; \mathbb{N} + 1 = (X + 1)\dot{\cup}(X' + 1)$
$\Rightarrow\; \mathbb{N} = (X + 1)\dot{\cup}(X' + 1)\dot{\cup}\{1\}$
$\Rightarrow\; (X + 1)' = (X' + 1) \cup \{1\}$

Applying this to the set $(X + 1)'$, we get

$$
\begin{aligned}
\omega(X) \;&=\; ((X+1)' + 1)' \\
&=\; ((X+1)'' + 1) \cup \{1\} \\
&=\; ((X+1) + 1) \cup \{1\} \\
&=\; (X + 2) \cup \{1\}.
\end{aligned}
$$

Thus, $X \subseteq \omega(X) \Longleftrightarrow X \setminus \{1\} \subseteq X + 2$, that is, whenever X contains an integer $n \geq 3$, it contains $n - 2$. Since $2 \notin (X + 2) \cup \{1\} \; \forall X \subseteq \mathbb{N}$, it follows that $X \subseteq \omega(X) \Longrightarrow X$ contains no even numbers. Hence, $\mathcal{F}$ consists of $\emptyset$ and sets of the form $F_n = \{2k - 1 \mid 1 \leq k \leq n\}$, along with their union $U = \{2k - 1 \mid k \in \mathbb{N}\}$, the set of all odd numbers.

6.18 Let I be any of the intervals $(c, d]$, $[c, d)$, $[c, d]$ and put $d - c = 3t$. Then $c + t < d - t$ and $(c + t, d - t) \subseteq I$. Clearly $I \subseteq (c - 1, d + 1)$, and then it follows from Exercise 6.12 that

$$
(a, b) \sim (c + t, c - t) \subseteq I \subseteq (c - 1, d + 1) \sim (a, b).
$$

Thus there are injections from (a, b) to I and back again, and so $(a, b) \sim I$ by Theorem 6.2.

6.19 Propose the definition

$$
\beta(1) = 1, \qquad \beta(\sigma(s)) = \tau(\beta(s)), \qquad s \in S;
$$

the set $X = \{s \in S \mid \beta(s) \text{ is defined}\}$ is then the whole of S, by axiom P3. So β is a map. Similarly, so is $\gamma : T \longrightarrow S$, given by

$$
\gamma(1) = 1, \qquad \gamma(\tau(s)) = \beta(\gamma(s)),
$$

and β, γ are clearly mutually inverse. So β is a bijection.

6.20 Let $|A| = a$, $|B| = b$ and $x \notin A \cup B$. Then put $\overline{A} = A \cup \{x\}$, $\overline{B} = B \cup \{x\}$, so that $|\overline{A}| = |\overline{B}|$. Then $\exists$ bijection $\beta : \overline{A} \longrightarrow \overline{B}$. Put

$$\beta(A) = C \qquad \text{and} \qquad \beta(x) = y,$$

so that $|A| = |C|$ and $\overline{B} = B \cup \{x\} = C \cup \{y\}$, $x \notin B$, $y \notin C$. If $x = y$, then $B = C$, and

$$b = |B| = |C| = |A| = a.$$

If $x \neq y$, consider $D = B \cap C = \overline{B} \backslash \{x, y\}$. Then $B = D \cup \{y\}$ and $C = D \cup \{x\}$, whence

$$b = |B| = |D| + 1 = |C| = |A| = a.$$

So we have deduced $a = b$ in both cases.

6.21 When a is finite, $a \geq a + 1 \Longrightarrow 0 \geq 1 > 0$, impossible. Let $|A| = a$ be infinite. By Theorem 6.3, there is an injection $\alpha : \mathbb{N} \longrightarrow A$. Then

$$\begin{aligned} a &= |\operatorname{Im}\alpha| + |A \backslash \operatorname{Im}\alpha| \\ &= |\mathbb{N}| + |A \backslash \operatorname{Im}\alpha|. \end{aligned}$$

Since

$$\begin{aligned} \beta : \quad \mathbb{N} &\longrightarrow \mathbb{N} \backslash \{1\} \\ n &\longmapsto n + 1 \end{aligned}$$

is a bijection, $1 + |\mathbb{N}| = |\mathbb{N}|$. Then we have

$$\begin{aligned} a &= |\mathbb{N}| + |A \backslash \operatorname{Im}\alpha| \\ &= 1 + |\mathbb{N}| + |A \backslash \operatorname{Im}\alpha| \\ &= 1 + a = a + 1. \end{aligned}$$

6.23 If $|A| = a$ is finite and $|B| = b$ is infinite, then Theorem 6.3 implies there are injections

$$\alpha : A \longrightarrow \mathbb{N}, \qquad \beta : \mathbb{N} \longrightarrow B,$$

whence $a \leq \aleph_0 \leq b$, so $a \leq b$ by transitivity. Since $a = b$ is impossible (by Theorem 6.3 again), $a < b$.

6.24 Since

$$\begin{aligned} \alpha : \quad \mathbb{N} &\longrightarrow \mathbb{N} \\ k &\longmapsto k + n \end{aligned}$$

is an injection,

$$\aleph_0 = |\mathbb{N}| = |\operatorname{Im}\alpha| + |\mathbb{N} \backslash \operatorname{Im}\alpha| = \aleph_0 + n.$$

6.26 This is a straightforward induction on n using Theorem 6.5.

6.28 Writing $\mathbb{Z} = \mathbb{N} \dot{\cup} \{0\} \dot{\cup} -\mathbb{N}$, we have

$$|\mathbb{Z}| = 2\aleph_0 + 1 = \aleph_0 + 1 = \aleph_0.$$

6.29 We have injections

$$
\begin{array}{ccc}
\mathbb{Z} & \longrightarrow & \mathbb{Q} \\
n & \longmapsto & n/1,
\end{array}
\qquad
\begin{array}{ccc}
\mathbb{Q} & \longrightarrow & \mathbb{Z} \times \mathbb{Z} \\
a/b & \longmapsto & (a,b),
\end{array}
$$

where a/b is in lowest terms. Thus,

$$\aleph_0 = |\mathbb{Z}| \leq |\mathbb{Q}| \leq |\mathbb{Z} \times \mathbb{Z}| = \aleph_0^2 = \aleph_0,$$

using Exercise 6.28. Then $|\mathbb{Q}| = \aleph_0$ by the Cantor–Schroeder–Bernstein theorem.

6.30 Given $\theta : \mathbb{N} \longrightarrow \mathbb{Z}/2\mathbb{Z}$, define $\chi^{-1}(\theta) = \{n \in \mathbb{N} \mid \theta(n) = 1\}$.

6.32 Given $r = .a_1 \ldots a_m \dot{b}_1 \ldots \dot{b}_n$, we can write

$$r = 2^{-m}a + 2^{-(n+m)}b \sum_{k \geq 0}(2^{-nk}),$$

where $a = a_1 \ldots a_m$, $b = b_1 \ldots b_n$ are ordinary binary numbers. Since the sum over k is just $2^n/(2^n - 1)$, r is rational, and clearly $0 = .\dot{0} \leq r \leq .\dot{1} = 1$. For the converse, let $q = a/b$, where a and b are integers and we can assume $0 < a < b$ to avoid triviality. In the process of (binary long) division of a by b, the successive remainders r_n, $n \in \mathbb{N}$, all satisfy $0 \leq r_n \leq b-1$ by Euclid's theorem. So we will find $m, n \in \mathbb{N}$ with $m \neq n$ and $a \leq m, n \leq a + b$ such that $r_m = r_n$. Moreover, for $n \geq a$, r_{n+1} is the remainder on dividing $2r_n$ by b. This implies that $r_{m+k} = r_{n+k}$ $\forall k \in \mathbb{N}$, and the binary expansion is recurring.

6.33 Since every cofinite set has a binary expansion that is recurring, they all correspond to rational numbers, and we already know that $|\mathbb{Q}| = \aleph_0$.

6.34 The bijection is the composite

$$
\begin{array}{ccccccc}
(-1,0) & \longrightarrow & (1,0) & \overset{\alpha}{\longrightarrow} & \mathbb{R}^+ & \longrightarrow & \mathbb{R}^-, \\
r & \longmapsto & -r & \longmapsto & \frac{-r}{1+r} & \longmapsto & \frac{r}{1+r}.
\end{array}
$$

6.35 The ordering on $[0,1)$ comes by ordering the sequences

$$\{\chi_A(n) \mid n \in \mathbb{N}\}, \qquad A \subseteq \mathcal{P}(\mathbb{N}), \qquad A \text{ non-cofinite,}$$

lexicographically. With the usual ordering of $\mathbb{Z}$, this extends to $\mathbb{R}$ via the obvious bijection $\beta : \mathbb{R} \longrightarrow \mathbb{Z} \times [0,1)$.

For the sum and product, take the termwise sum and product of Cauchy sequences (as in (6.26)) in $[0, 1)$ and extend to the whole of $\mathbb{R}$ via β and the usual sum and product on $\mathbb{Z}$. [N.B. These are just the *definitions*: it takes quite a lot of work to prove that they are correct.]

6.37 We have

$$c \le b + c \le c + c = 2c \le c,$$

by the previous exercise. So $b + c = c$ by the Cantor–Schroeder–Bernstein theorem.

Guide to the Literature

The references for Chapter 1 are [9] and [6]. The former is a more detailed and practical study of the structure of proofs and includes a large number of examples in its leisurely approach. The latter takes further the illustrations from number theory used in this chapter and forms the classic introduction to this subject.

The interface between logic and mathematics, which is hinted at in Chapter 2 and forms the chief motivation for this book, is extended in [10]. A fuller treatment of the syllogistic is to be found in [3].

The first half of the classic text [7] gives a rather more condensed treatment of Chapter 3 and the next two, and goes on to introduce the basic ideas of group theory. Elementary set theory occupies the early chapters in [4], and [10] contains a full and rigorous introduction. The study of topology is introduced and carried to some depth in [1].

The basic material of Chapter 4 is covered in [8] as well as in [7], and the former also contains an introduction to abstract algebra and the construction of the real numbers by Dedekind section.

The key reference for Chapter 5 is again [7], and Peano's axioms are treated in [8].

The material in Chapter 6 comes directly from its source [2]. The more recent reference [5], which is destined to become a classic, gives an accessible and entertaining account of a great diversity of numerical curiosities, including the very large and very small surreal numbers in its final chapter. The definitive treatment of the continuum hypothesis is to be found in [4].

Some of these books may be out of print, although copies of all of them are held at the University of Nottingham library. It is hoped that other volumes yet to be published in this Springer series will also provide further reading on most of the topics introduced in this book.

Bibliography

[1] M. A Armstrong, *Basic Topology*, McGraw-Hill, London 1979.

[2] G. Cantor, *Contributions to the Founding of the Theory of Transfinite Numbers*, Dover, New York 1955.

[3] M. Clark, *The Place of Syllogistic in Logical Theory*, Nottingham University Press, Nottingham 1980.

[4] P. J. Cohen, *Set Theory and the Continuum Hypothesis*, Benjamin, New York 1966.

[5] J. H. Conway and R. K. Guy, *The Book of Numbers*, Copernicus, New York 1996.

[6] H. Davenport, *The Higher Arithmetic: An Introduction to the Theory of Numbers*, 4th edn, Hutchinson, London 1970.

[7] J. A. Green, *Sets and Groups: A First Course in Algebra*, Chapman-Hall, London 1995.

[8] F. D. Parker, *The Structure of Number Systems*, Prentice-Hall, Engelwood 1966.

[9] L. Solow, *How to Read and Do Proofs*, Wiley, New York 1990.

[10] R. R. Stoll, *Sets, Logic and Axiomatic Theories*, 2nd edn, Freeman, San Francisco 1974.

Dramatis Personae
(in approximate order of appearance)

The use of symbols in mathematics is of fundamental importance. Not only do they provide convenient abbreviations, but correct choice of notation can often play a crucial role in solving a problem. And conversely: if you don't believe me, try dividing XCVIII by XLIV. Just as in everyday life a word should ideally stand for a single concept, so in mathematics should a symbol represent a unique quantity. But in both cases tradition dictates otherwise, and it is easy to think of words and symbols that can have two or more meanings; there are at least three glaring examples of the latter in the foregoing, namely, $[\pi]$, θ^{-1} and A'. Again in both cases, the problem of ambiguity is solved by taking account of the context, with a brief explanatory note if necessary. Thus, when we write "let A, A' be two arbitrary subsets of S", we are not saying that A' is the complement of A in S, just as when we say "Jimmy Conners is leading by one set to love" we do not mean one set of saucepans.

References are given to the relevant chapters in bold lettering

$\mathbb{N}$	the (set of) positive integers (Chapter **1**)
$\mathbb{Z}$	the integers (**1**)
$\mathbb{Q}$	the rational numbers (**1**, **4**)
$\mathbb{R}$	the real numbers (**1**, **4**)
$\mathbb{C}$	the complex numbers (**1**, **4**)
$\mathbb{H}$	the quaternions (**4**)
$\mathbb{A}$	the Cayley numbers (**4**)
$\mathbb{Z}/n\mathbb{Z}$	the integers modulo n (**4**)
$\mathbb{R}^+$	the positive reals (**6**)
$\mathbb{R}^-$	the negative reals (**6**)

$\mathbb{R}^n$	real Euclidean space of dimension n (**3**)
0	zero, the additive identity (**1**)
1	one, or unity, the multiplicative identy (**1**)
$\aleph_0$	the cardinality of $\mathbb{N}$ (**3**, **6**)
c	the cardinal of the continuum (**3**, **6**)
$n!$	n factorial (**1**)
$\sqrt{n}$	root n (**1**)
$\sqrt[3]{n}$	cube-root n (**1**)
$\binom{n}{k}$	n choose k (**1**)
$\min(m, n)$	the smaller of m and n (**1**)
$\max(m, n)$	the larger of m and n (**1**)
$-a$	additive inverse of a (**1**)
a^{-1}	multiplicative inverse of a (**1**)
x^n	nth power of x (**1**)
$\lvert x \rvert$	modulus of x (**3**)
$[x]$	integer part of x (**4**)
$=$	is equal to (**1**, **3**, $\ldots$)
$<$	is less than (**1**, **4**)
$\leq$	is less than or equal to (**1**, **4**)
$\nleq, >$	is greater than (**1**, **4**)
$\nless, \geq$	is greater than or equal to (**1**, **4**)
$\mid$	divides (**1**, **4**)
$\nmid$	does not divide (**1**)
$a \equiv b \pmod{n}$	n divides $a - b$ (**3**, **4**)
$.a_1 a_2 \ldots$	binary (**6**)
$.a_1 \dot{a}_2$	recurring binary (**6**)
(a, b)	highest common factor of a and b (**1**)
$[a, b]$	least common multiple of a and b (**1**)
$[a, b)$	half-open interval (**3**, **6**)
(x, y)	point in Cartesian plane (**3**)
	ordered pair (**3**)
$(x, y, \ldots)$	ordered set, or tuple (**3**)
$\{x, y, \ldots\}$	(unordered) set (**3**)
$\{x \mid \ldots\}$	subset (**3**)
$[a]$	equivalence class of a (**4**)
$[a, b]$	equivalence class of (a, b) (**4**)
(a_n)	sequence of rational numbers (**4**)
$[a_n]$	equivalence class of (a_n) (**4**)
IH	inductive hypothesis (**1**)
$\sim P$	negation of P (**1**, **2**)
$\wedge$	and (**2**)

$\vee$	or (**2**)
$\triangle$	either or but not both (**2**)
$\Rightarrow$	implies (**1**, **2**, ...)
$\Leftrightarrow$	if and only if (**1**, ...)
$\equiv$	is equivalent to (**2**)
	congruence (**3**, **4**)
T	true (**2**)
F	false (**2**)
$\forall$	for all (each, every, any) (**2**, ...)
$\exists$	there is (exists) (**2**, ...)
$\exists\vert$	there is exactly one (**3**)
$\in$	is a member of (**3**)
$\notin$	is not a member of (**3**)
$\subseteq$	is contained in (**3**)
$\not\subseteq$	is not contained in (**3**)
$\subset$	is properly contained in (**3**)
$\emptyset$	is the empty set (**3**)
$\vert S\vert$, $\#S$	the cardinality of S (**3**, **6**)
A'	the complement of A (**3**)
$\cup$	union (**3**)
$\cap$	intersection (**3**)
$\setminus$	difference (**3**)
$\triangle$	symmetric difference (**3**)
$\times$	Cartasian product (**3**)
$A^{\times n}$, A^n	nth Cartasian power (**3**)
$\mathcal{P}(S)$	the power set of S (**3**)
$\equiv$, $\sim$	(equivalence) relation (**4**)
(R)	reflexive law (**5**)
(S)	symmetry law (**5**)
(T)	transitive law (**5**)
(I)	irreflexive law (**5**)
(O)	ordering law (**5**)
(L)	total (ordering) (**5**)
(W)	well (ordering) (**5**)
(A, B)	Dedekind section (**4**)
$\sum$	(iterated) summation (**1**)
$\prod$	product (**1**)
$\bigwedge$	conjunction (**2**)
$\bigvee$	disjunction (**2**)
$\bigcap$	union (**3**)
$\bigcup$	intersection (**3**)

$\theta : A \to B$	a map θ from A to B (5)
$a \mapsto b$	a is mapped to b (5)
$\theta(a)$	the value of θ at a (5)
$\mathrm{Im}\,\theta$	the image of θ (5)
$\mathrm{Map}(A, B)$	the set of maps from A to B (5)
θ^{-1}	the inverse of θ (5)
ι, inc	inclusion (5)
ν, nat	natural map (5)
δ	diagonal map
$A/\!\sim$	quotient set (5)
$\theta \times \phi$	product of maps (5)
χ_A	characteristic function of A (5)
ϕ	Euler totient function (5)
$\mathrm{Sym}(A)$	symmetric group on A (5)
S_n	symmetric group of degree n (5)

Index

abuse of notation, 60, 93
addition, 106, 115
Adian, S.I., 18
algebra
– fundamental theorem of, 82
– linear, 60
algebraic closure, 82
alphabet, 87
analysis, 68
Aristotle, 43
arithmetic
– fundamental laws of, 3
– fundamental theorem of, 31
– modular, 77
Axiom of Choice, 68, 86

Bernstein, 118
bijection, 74, 102
binary point, 126
binomial
– coefficients, 21
– theorem, 21
Boole, 37
Burnside, 18

calculus, 98
– predicate, 43
– propositional, 36
Cantor, 114, 118
cardinal, 113
– number, 115
– of the continuum, 56, 128
cardinality, 56, 113
Cartesian
– plane, 57, 60

– product, 60, 123
Cayley, 83
chain, 122
closed, 70
– formula, 21
– set, 70
codomain, 90
coefficient, 96
Cohen, 130
complement, 55
completeness, 84
conclusion, 43
condition
– necessary, 8
– sufficient, 8
congruence, 75
conjunction, 36
continuum hypothesis, 130
contradiction, 5, 141
contrapositive, 8
converse, 8
coprime, 28
coset, 78
counter-example, 46, 48

Dedekind
– property, 83
– sections, 83
degree, 96
denominator, 6
diagram
– Lewis Carroll, 45
– Venn, 60
difference, 59
– symmetric, 59

disjoint, 60
disjunction, 36
divisors of zero, 4
domain, 90

element, 54
empty
– product, 14
– set, 56
equivalence
– class, 73
– relation, 73
Euclidean
– algorithm, 29
– geometry, 5
– theorem, 6, 28, 82
Euler totient function, 104
exponentiation, 115

family, 53, 66
field, 4, 82
figure, 44
fraction, 6
function
– characteristic, 99
– elementary, 98
– exponential, 98
– trigonometric, 98

graph, 90
group, 63

Hamilton, 83
highest common factor, 28
horse, 18

identity, 141
– additive, 4, 82
– map, 92
– multiplicative, 4, 82
implication, 8, 36
inclusion, 94
indeterminate, 96
induction
– cumulative, 13
– double, 14
– mathematical, 11
– simultaneous, 17
inductive
– base, 11
– definition, 19
– hypothesis, 11
– proof, 10
– step, 11

– variable, 11
injection, 99
integer, 80
– part, 75
– positive, 1
integral domain, 4, 80
intersection, 59
interval
– half-open, 67
– open, 68
inverse, 98, 102
– additive, 4
– map, 98
– multiplicative, 4

law, 3, 62
– associative, 3, 115
– cancellation, 107, 110, 111
– commutative, 3, 115
– de Morgan, 40
– distributive, 3, 115
– idempotent, 48, 65, 148
– of exportation, 42
– of importation, 42
– of indices, 115
– of permutation, 42
– of syllogism, 42
– of trichotomy, 86
– universal, 115
lemma, 105
length, 87
limit, 84
logic, 35
– formal, 36
– of quantifiers, 43
– symbolic, 36
logical
– connective, 36
– constant, 36
– equivalence, 8, 37
– variable, 36
lowest
– common multiple, 33
– terms, 6

map, 89
– argument of, 90
– bijective, 102
– codomain of, 90
– composite, 91, 114
– constant, 97
– diagonal, 94
– domain of, 19, 90

– extension of, 94
– image of, 90
– inclusion, 94
– injective, 99
– inverse, 98
– natural, 95
– one-to-one, 99
– onto, 100
– polynomial, 96
– pre-image of, 92
– projection, 95
– restriction, 94
– substitution, 96
– surjective, 100
– value of, 90
member, 54
modus ponens, 41
modus tollens, 41
multiplication, 107, 115

negation, 5, 36, 141
Notts County, 36
numbers, 1
– Catalan, 5, 24
– Cayley, 83
– complex, 1, 82
– Fibonacci, 23
– irrational, 6
– rational, 1, 5, 81
– real, 1, 127
numerator, 6

Ol'shanskii, A. Yu, 18
operation, 3
– binary, 37, 58, 95
– logical, 36, 58
– set-theoretical, 58
– unary, 37
ordered pair, 60
ordering, 108
– lexicographic, 87
– partial, 85
– quasi, 86
– shortlex, 87
– total, 85
– well, 85, 110

parrot, 43
partition, 66
Pascal's Triangle, 21
Peano's Axioms, 105
permutation, 103
pigeon-hole principle, 102

point, 54
pointwise
– addition, 97
– multiplication, 97
– operation, 97
– product, 97
– sum, 97
polynomial, 72
postulate, 105, 115
power, 20
– set, 65
predicate, 43
premise, 43
– major, 43
– minor, 43
prime, 9
product, 99
progression
– arithmetic, 1
– geometric, 18
proof, 1
– by contradiction, 5
– by contraposition, 9
– by counter-example, 46
– by induction, 1, 10
– by reduction to absurdity, 5
proposition, 35
– categorical, 43
Pythagoras' theorem, 6

quantifier
– existential, 48
– universal, 48
quaternions, 83

relation, 37, 71
– reflexive, 72
– symmetric, 72
– transitive, 72
relatively prime, 28
residue, 67
– class, 67
ring
– commutative, 4
– with identity, 4
rules of indices, 20
Russell's paradox, 114

Schroeder, 118
semantics, 36
sequence, 2, 19, 127
– Cauchy, 84
– convergent, 84

– null, 84
set, 53
– closed, 70
– countable, 121
– countably infinite, 56, 121
– equipotent, 114
– equivalent, 114
– finite, 56, 101, 113, 121
– infinite, 56, 113
– open, 68
– power, 65
– quotient, 95
– similar, 114
– uncountable, 121, 126
singleton, 56
square-root, 92, 154
subgroup, 78
subject, 43
subset, 54, 70
– diagonal, 94
surjection, 100
syllogism, 43
symmetric group, 103
symmetry, 21
syntax, 36

tautology, 141

term, 43
– major, 43
– middle, 43
– minor, 43
Theophrastus, 44
topological space, 69
topology, 69
– cofinite, 70
– discrete, 149
transversal, 68
truth
– table, 39
– value, 36

uncountable, 121
union, 59
unique, 6

variable, 50
– bound, 50
– free, 50

well-ordering principle, 27
word, 87

Zermelo, 121